STUDENT'S SOLUTIONS MANUAL

COLLEGE ALGEBRA

THIRD EDITION

JEROME E. KAUFMANN

PWS Publishing Company
Boston

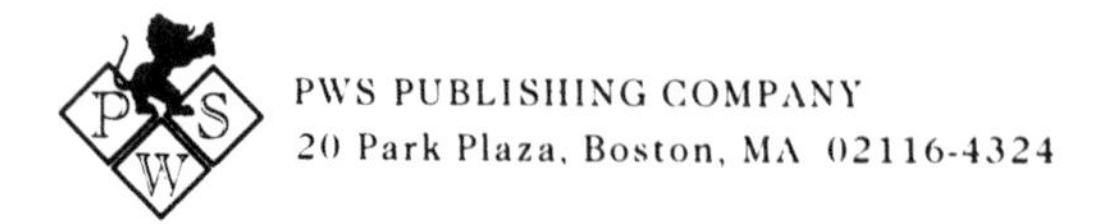

I(T)P™
International Thomson Publishing
The trademark ITP is used under license

PWS Publishing Company is a division of Wadsworth, Inc.

Printed in the United States of America by Malloy Lithographing

1 2 3 4 5 6 7 8 9 10 -- 98 97 96 95 94 93

ISBN 0-534-93503-6

PREFACE

This student supplement has been prepared to accompany the third edition of College Algebra. Detailed solutions for approximately one-fourth of the exercises are included. In general, we have included the solutions for problems numgered 1, 5, 9, 13, and so on. If such problems are short-answer problems with no work to be shown, however, they have occasionally not been included since the answers are in the back of the book.

Certainly many of the exercises can be solved in different ways. In order to keep this supplement a reasonable size, we have shown only one approach to a problem. Therefore, you may use an approach which is different and perhaps simpler than the way shown. However, if you cannot solve a problem, then referring to our suggestions should be of some help. It is to your advantage to use this supplement only after you have made a sincere effort to solve the problem by yourself.

Good luck in your study of college algebra.

J. Kaufmann

CONTENTS

CHAPTER 1

Problem Set 1.1

61. $-3ab-2c = -3(-4)(7)-2(-8) = 84+16 = 100$

65. $\dfrac{-2x+7y}{x-y} = \dfrac{-2(-3)+7(-2)}{-3-(-2)} = \dfrac{6-14}{-1} = \dfrac{-8}{-1} = 8$

69. $5x+4y-9y-2y = 5x-7y = 5(2)-7(-8) = 10+56 = 66$

73. $|3x+y|+|2x-4y| = |3(5)+(-3)|+|2(5)-4(-3)| = |12|+|22| = 12+22 = 34$

77. $2(3x+4)-3(2x-1) = 6x+8-6x+3 = 11$

Problem Set 1.2

1. $2^{-3} = \dfrac{1}{2^3} = \dfrac{1}{8}$

5. $\dfrac{1}{3^{-3}} = \dfrac{1}{\dfrac{1}{3^3}} = \dfrac{1}{\dfrac{1}{27}} = 27$

9. $\left(-\dfrac{2}{3}\right)^{-3} = \dfrac{1}{\left(-\dfrac{2}{3}\right)^3} = \dfrac{1}{-\dfrac{8}{27}} = -\dfrac{27}{8}$

13. $\dfrac{1}{\left(\dfrac{4}{5}\right)^{-2}} = \left(\dfrac{4}{5}\right)^2 = \dfrac{16}{25}$

17. $10^{-6}\cdot 10^4 = 10^{-2} = \dfrac{1}{10^2} = \dfrac{1}{100}$

21. $\left(3^{-2}\right)^{-2} = 3^4 = 81$

25. $\left(3^{-1}\cdot 2^2\right)^{-1} = 3^1\cdot 2^{-2} = \dfrac{3}{2^2} = \dfrac{3}{4}$

29. $\left(\dfrac{2^{-2}}{5^{-1}}\right)^{-2} = \dfrac{2^4}{5^2} = \dfrac{16}{25}$

33. $\dfrac{2^3}{2^{-3}} = 2^{3-(-3)} = 2^6 = 64$

37. $3^{-2}+2^{-3} = \dfrac{1}{3^2}+\dfrac{1}{2^3} = \dfrac{1}{9}+\dfrac{1}{8} = \dfrac{8}{72}+\dfrac{9}{72} = \dfrac{17}{72}$

41. $\left(2^{-4}+3^{-1}\right)^{-1} = \left(\dfrac{1}{2^4}+\dfrac{1}{3^1}\right)^{-1} = \left(\dfrac{1}{16}+\dfrac{1}{3}\right)^{-1} = \left(\dfrac{3}{48}+\dfrac{16}{48}\right)^{-1}$

$= \left(\dfrac{19}{48}\right)^{-1} = \dfrac{1}{\left(\dfrac{19}{48}\right)^1} = \dfrac{48}{19}$

45. $a^2\cdot a^{-3}\cdot a^{-1} = a^{2+(-3)+(-1)} = a^{-2} = \dfrac{1}{a^2}$

49. $\left(x^3y^{-4}\right)^{-1} = x^{-3}y^4 = \dfrac{y^4}{x^3}$

53. $\left(2x^2y^{-1}\right)^{-2} = 2^{-2}x^{-4}y^2 = \dfrac{y^2}{2^2\cdot x^4} = \dfrac{y^2}{4x^4}$

57. $\left(\dfrac{2a^{-1}}{3b^{-2}}\right)^{-2} = \dfrac{2^{-2}a^2}{3^{-2}b^4} = \dfrac{3^2\cdot a^2}{2^2\cdot b^4} = \dfrac{9a^2}{4b^4}$

61. $\dfrac{a^2b^{-3}}{a^{-1}b^{-2}} = a^{2-(-1)}b^{-3-(-2)} = a^3b^{-1} = \dfrac{a^3}{b}$

65. $(-3xy^3)^3 = (-3)^3(x)^3(y^3)^3 = -27x^3y^9$

69. $\frac{-72x^8}{-9x^2} = -8x^{8-2} = -8x^6$

73. $(-6a^5y^{-4})(-a^{-7}y) = (-6)(-1)a^{5+(-7)}y^{-4+1} = 6a^{-2}y^{-3} = \frac{6}{a^2y^3}$

77. $\frac{-35a^3b^{-2}}{7a^5b^{-1}} = -5a^{3-5}b^{-2-(-1)} = -5a^{-2}b^{-1} = -\frac{5}{a^2b}$

81. $x^{-1}+x^{-2} = \frac{1}{x} + \frac{1}{x^2} = \frac{x+1}{x^2}$

85. $3a^{-2}+2b^{-3} = \frac{3}{a^2} + \frac{2}{b^3} = \frac{3b^3+2a^2}{a^2b^3}$

89. $(3x^a)(4x^{2a+1}) = 12x^{a+2a+1} = 12x^{3a+1}$

93. $\frac{x^{3a}}{x^a} = x^{3a-a} = x^{2a}$

97. $\frac{(xy)^b}{y^b} = \frac{x^by^b}{y^b} = x^b$

109. $\sqrt{900000000} = \sqrt{9\cdot10^8} = \sqrt{9}\sqrt{10^8} = 3\cdot10^4 = 30000$

Problem Set 1.3

1. $(5x^2-7x-2)+(9x^2+8x-4) = (5+9)x^2+(-7+8)x+(-2+(-4)) = 14x^2+x-6$

5. $(3x-4)-(6x+3)+(9x-4) = 3x-4-6x-3+9x-4 = 3x-6x+9x-4-3-4 = 6x-11$

9. $5(x-2)-4(x+3)-2(x+6) = 5x-10-4x-12-2x-12 = 5x-4x-2x-10-12-12 = -x-34$

13. $6a^3b^2(5ab-4a^2b+3ab^2) = 6a^3b^2(5ab)-(6a^3b^2)(4a^2b)+(6a^3b^2)(3ab^2)$
$= 30a^4b^3-25a^5b^3+18a^4b^4$

17. $(n-4)(n-12) = n^2-12n-4n+48 = n^2-16n+48$

21. $(3x-1)(2x+3) = 6x^2+9x-2x-3 = 6x^2+7x-3$

25. $(x+4)^2 = x^2+2(4x)+16 = x^2+8x+16$

29. $(x+2)(x-4)(x+3) = (x+2)(x^2-x-12)$
$= x(x^2)-x(x)-x(12)+2(x^2)-2(x)-2(12) = x^3+x^2-14x-24$

33. $(x-1)(x^2+3x-4) = x(x^2)+x(3x)-x(4)-1(x^2)-1(3x)-1(-4)$
$= x^3+3x^2-4x-x^2-3x+4 = x^3+2x^2-7x+4$

37. $(3x+2)(2x^2-x-1) = 3x(2x^2)-3x(x)-(3x)(1)+2(2x^2)-2(x)-2(1)$
$= 6x^3-3x^2-3x+4x^2-2x-2 = 6x^3+x^2-5x-2$

For Problems 41 and 45, use the pattern $(a+b)(a-b) = a^2-b^2$.

41. $(5x-2)(5x+2) = (5x)^2-(2)^2 = 25x^2-4$

45. $(2x+3y)(2x-3y) = (2x)^2-(3y)^2 = 4x^2-9y^2$

49. Use the pattern $(a+b)^3 = a^3+3a^2b+3ab^2+b^3$.
$(2x+1)^3 = (2x)^3+3(2x)^2(1)+3(2x)(1)^2+(1)^3 = 8x^3+12x^2+6x+1$

53. Use the pattern $(a-b)^3 = a^3-3a^2b+3ab^2-b^3$.
$(5x-2y)^3 = (5x)^3-3(5x)^2(2y)+3(5x)(2y)^2-(2y)^3 = 125x^3-150x^2y+60xy^2-8y^3$

For Problems 57, 61, and 65 use the binomial expansion pattern.

57. $(x-y)^5 = x^5+5(x)^4(-y)+10(x)^3(-y)^2+10(x)^2(-y)^3+5(x)(-y)^4+(-y)^5$
$= x^5-5x^4y+10x^3y^2-10x^2y^3+5xy^4-y^5$

61. $(2a-b)^6 = (2a)^6+6(2a)^5(-b)^1+15(2a)^4(-b)^2+20(2a)^3(-b)^3$
$+15(2a)^2(-b)^4+6(2a)(-b)^5+(-b)^6$
$= 64a^6-192a^5b+240a^4b^2-160a^3b^3+60a^2b^4-12ab^5+b^6$

65. $(2a-3b)^5 = (2a)^5+5(2a)^4(-3b)+10(2a)^3(-3b)^2+10(2a)^2(-3b)^3+5(2a)(-3b)^4+(-3b)^5$
$= 32a^5-240a^4b+720a^3b^2-1080a^2b^3+810ab^4-243b^5$

69. $\dfrac{30a^5-24a^3+54a^2}{-6a} = \dfrac{30a^5}{-6a} - \dfrac{24a^3}{-6a} + \dfrac{54a^2}{-6a} = -5a^4+4a^2-9a$

73. $(x^a+y^b)(x^a-y^b) = (x^a)^2-(y^b)^2 = x^{2a}-y^{2b}$

77. $(2x^b-1)(3x^b+2) = 6x^{2b}+2x^b(2)-1(3x^b)-1(2) = 6x^{2b}+4x^b-3x^b-2 = 6x^{2b}+x^b-2$

81. $(x^a-2)^3 = (x^a)^3-3(x^a)^2(2)+3(x^a)(2)^2-(2)^3 = x^{3a}-6x^{2a}+12x^a-8$

Problem Set 1.4

1. $6xy-8xy^2 = 2xy(3)-2xy(4y) = 2xy(3-4y)$

5. $3x+3y+ax+ay = 3(x+y)+a(x+y) = (x+y)(3+a)$

9. $9x^2-25 = (3x)^2-(5)^2 = (3x-5)(3x+5)$

13. $(x+4)^2-y^2 = [(x+4)+y][(x+4)-y] = (x+4+y)(x+4-y)$

17. We need two integers whose product is -14 and whose sum is -5. They are -7 and 2. Therefore,
$$x^2-5x-14 = (x-7)(x+2).$$

21. We need two integers whose product is -36 and whose sum is 7. The pairs of factors of -36 are (1)(-36),(-1)(36),(2)(-18),(-2)(18),(3)(-12), (-3)(12), (4)(-9),(-4)(9), and (-6)(6). Since none of these produces a sum of 7, the expression $x^2+7x-36$ is <u>not factorable</u> using integers.

25. $10x^2-33x-7$ → sum of -33

product of 10(-7) = -70

We need two integers whose product is -70 and whose sum is -33. They are -35 and 2. Therefore,
$$10x^2-33x-7 = 10x^2-35x+2x-7 = 5x(2x-7)+1(2x-7) = (2x-7)(5x+1).$$

29. Use the pattern $a^3+b^3 = (a+b)(a^2-ab+b^2)$.
$$64x^3+27y^3 = (4x)^3+(3y)^3 = (4x+3y)(16x^2-12xy+9y^2).$$

33. $x^3-9x = x(x^2-9) = x(x+3)(x-3)$

37. $2n^3+6n^2+10n = 2n(n^2+3n+5)$

The expression n^2+3n+5 will not factor since there are no two integers whose product is 5 and whose sum is 3.

41. This is of the form $(a-b)^2 = a^2-2ab+b^2$. Therefore,
$$36a^2-12a+1 = (6a-1)^2.$$

45. $2n^2-n-5$ → sum of -1

↓

product of $2(-5) = -10$

We need two integers whose product is -10 and whose sum is -1. The pairs of factors of -10 are (1)(-10),(-1)(10),(2)(-5), and (-2)(5). None of these produce a sum of -1; thus. the given expression is not factorable.

49. $4x^3+32 = 4(x^3+8)$

Now x^3+8 can be treated as the sum of two cubes.

$4x^3+32 = 4(x^3+8) = 4(x+2)(x^2-2x+4)$

53. $2x^4y-26x^2y-96y = 2y(x^4-13x^2-48) = 2y(x^2-16)(x^2+3) = 2y(x+4)(x-4)(x^2+3)$

57. $x^2+8x+16-y^2 = (x+4)^2-(y)^2 = (x+4+y)(x+4-y)$

61. $60x^2-32x-15$ → sum of -32

↓

product of $60(-15) = -900$

We need two integers whose sum is -32 and whose product is -900. They are -50 and 18. (If you have trouble finding these integers, read the discussion in Problem 75 in the text.)

$$60x^2-32x-15 = 60x^2-50x+18x-15 = 10x(6x-5)+3(6x-5) = (6x-5)(10x+3).$$

65. Use the difference of two squares pattern.

$$x^{2a}-16 = (x^a)^2-(4)^2 = (x^a+4)(x^a-4)$$

69. We need two integers whose product is -28 and whose sum is -3. They are -7 and 4. Therefore,

$$x^{2a}-3x^a-28 = x^{2a}-7x^a+4x^a-28 = x^a(x^a-7)+4(x^a-7) = (x^a-7)(x^a+4).$$

73. Use the difference of two squares pattern twice.

$$x^{4n}-y^{4n} = (x^{2n})^2-(y^{2n})^2 = (x^{2n}-y^{2n})(x^{2n}+y^{2n})$$

$$= [(x^n)^2-(y^n)^2](x^{2n}+y^{2n}) = (x^n-y^n)(x^n+y^n)(x^{2n}+y^{2n})$$

Problem Set 1.5

1. $\frac{14x^2y}{21xy} = \frac{\cancel{7}\cdot 2\cdot \cancel{x^2}^{\,x}\cdot \cancel{y}}{\cancel{7}\cdot 3\cdot \cancel{x}\cdot \cancel{y}} = \frac{2x}{3}$

5. $\frac{a^2+7a+12}{a^2-6a-27} = \frac{(a+4)\cancel{(a+3)}}{(a-9)\cancel{(a+3)}} = \frac{a+4}{a-9}$

9. $\frac{x^3-y^3}{x^2+xy-2y^2} = \frac{\cancel{(x-y)}(x^2+xy+y^2)}{\cancel{(x-y)}(x+2y)} = \frac{x^2+xy+y^2}{x+2y}$

13. $\frac{4x^2}{5y^2}\cdot\frac{15xy}{24x^2y^2} = \frac{\cancel{4}\cdot \cancel{15}^{\,3}\cdot \cancel{x^3}^{\,x}\cdot \cancel{y}}{\cancel{5}\cdot \cancel{24}_{\,\cancel{6}\,2}\cdot \cancel{x^2}\cdot \cancel{y^4}_{\,y^3}} = \frac{x}{2y^3}$

17. $\frac{7a^2b}{9ab^3} \div \frac{3a^4}{2a^2b^2} = \frac{7a^2b}{9ab^3}\cdot\frac{2a^2b^2}{3a^4} = \frac{7\cdot 2\cdot \cancel{a^4}\cdot \cancel{b^3}}{9\cdot 3\cdot \cancel{a^5}_{\,a}\cancel{b^3}} = \frac{14}{27a}$

21. $\frac{5a^2+20a}{a^3-2a^2} \cdot \frac{a^2-a-12}{a^2-16} = \frac{5a(a+4)(a-4)(a+3)}{a^2(a-2)(a+4)(a-4)} = \frac{5(a+3)}{a(a-2)}$

25. $\frac{9y^2}{x^2+12x+36} \div \frac{12y}{x^2+6x} = \frac{9y^2}{x^2+12x+36} \cdot \frac{x^2+6x}{12y} = \frac{(9y^2)(x)(x+6)}{(12y)(x+6)(x+6)} = \frac{3xy}{4(x+6)}$

29. $\frac{x+4}{6} + \frac{2x-1}{4} = (\frac{x+4}{6})(\frac{2}{2})+(\frac{2x-1}{4})(\frac{3}{3}) = \frac{2x+8}{12} + \frac{6x-3}{12} = \frac{2x+8+6x-3}{12} = \frac{8x+5}{12}$

33. $\frac{7}{16a^2b} + \frac{3a}{20b^2} = (\frac{7}{16a^2b})(\frac{5b}{5b}) + (\frac{3a}{20b^2})(\frac{4a^2}{4a^2}) = \frac{35b}{80a^2b^2} + \frac{12a^3}{80a^2b^2} = \frac{35b+12a^3}{80a^2b^2}$

37. $\frac{3}{4x} + \frac{2}{3y} - 1 = (\frac{3}{4x})(\frac{3y}{3y}) + (\frac{2}{3y})(\frac{4x}{4x}) - 1(\frac{12xy}{12xy}) = \frac{9y}{12xy} + \frac{8x}{12xy} - \frac{12xy}{12xy} = \frac{9y+8x-12xy}{12xy}$

41. $\frac{4x}{x(x+7)} + \frac{3}{x} = \frac{4x}{x(x+7)} + (\frac{3}{x})(\frac{x+7}{x+7}) = \frac{4x}{x(x+7)} + \frac{3x+21}{x(x+7)} = \frac{4x+3x+21}{x(x+7)} = \frac{7x+21}{x(x+7)}$

45. $\frac{3}{x+1} + \frac{x+5}{(x+1)(x-1)} - \frac{3}{x-1} = \frac{3(x-1)+x+5-3(x+1)}{(x+1)(x-1)} = \frac{3x-3+x+5-3x-3}{(x+1)(x-1)} = \frac{x-1}{(x+1)(x-1)}$

$= \frac{1}{x+1}$

49. $\frac{5}{x^2-1} - \frac{2}{x^2+6x-16} = \frac{5}{(x+1)(x-1)} - \frac{2}{(x+8)(x-2)}$

$= \frac{5(x+8)(x-2)-2(x+1)(x-1)}{(x+1)(x-1)(x+8)(x-2)} = \frac{5x^2+30x-80-2x^2+2}{(x+1)(x-1)(x+8)(x-2)}$

$= \frac{3x^2+30x-78}{(x+1)(x-1)(x+8)(x-2)}$

53. $\frac{2n^2}{n^4-16} - \frac{n}{n^2-4} + \frac{1}{n+2} = \frac{2n^2}{(n^2+4)(n+2)(n-2)} - \frac{n}{(n+2)(n-2)} + \frac{1}{n+2}$

$= \frac{2n^2-n(n^2+4)+1(n^2+4)(n-2)}{(n^2+4)(n+2)(n-2)}$

$= \frac{2n^2-n^3-4n+n^3-2n^2+4n-8}{(n^2+4)(n+2)(n-2)} = \frac{-8}{(n^2+4)(n+2)(n-2)}$

57. (a) $\frac{7}{x-1} + \frac{2}{1-x} = \frac{7}{x-1} - \frac{2}{x-1} = \frac{7-2}{x-1} = \frac{5}{x-1}$

61. $\frac{1+\frac{1}{x}}{1-\frac{1}{x}} = \left(\frac{1+\frac{1}{x}}{1-\frac{1}{x}}\right)(\frac{x}{x}) = \frac{x(1)+x(\frac{1}{x})}{x(1)-x(\frac{1}{x})} = \frac{x+1}{x-1}$

65. $\left(\frac{\frac{-2}{x} - \frac{4}{x+2}}{\frac{3}{x(x+2)} + \frac{3}{x}}\right)\left(\frac{x(x+2)}{x(x+2)}\right) = \frac{-2(x+2)-4x}{3+3(x+2)} = \frac{-2x-4-4x}{3+3x+6} = \frac{-6x-4}{3x+9}$

69. $\frac{a}{\frac{1}{a}+4} + 1 = \frac{a}{\frac{1+4a}{a}} + 1 = \frac{a^2}{1+4a} + 1 = \frac{a^2+4a+1}{4a+1}$

73. $$\frac{\frac{1}{x+h+1}-\frac{1}{x+1}}{h} = \left[\frac{(x+1)(x+h+1)}{(x+1)(x+h+1)}\right]\left[\frac{\frac{1}{x+h+1}-\frac{1}{x+1}}{h}\right]$$

$$= \frac{x+1-(x+h+1)}{h(x+1)(x+h+1)} = \frac{x+1-x-h-1}{h(x+1)(x+h+1)}$$

$$= \frac{-h}{h(x+1)(x+h+1)} = -\frac{1}{(x+1)(x+h+1)}$$

77. $$\frac{x^{-1}+y^{-1}}{x-y} = \frac{\frac{1}{x}+\frac{1}{y}}{x-y} = \left(\frac{xy}{xy}\right)\left(\frac{\frac{1}{x}+\frac{1}{y}}{x-y}\right) = \frac{(xy)\left(\frac{1}{x}\right)+(xy)\left(\frac{1}{y}\right)}{(xy)(x-y)} = \frac{y+x}{x^2y-xy^2}$$

Problem Set 1.6

1. $\sqrt{81}$ means the principal or nonnegative square root of 81. Therefore, $\sqrt{81} = 9$.

5. $\sqrt{\frac{36}{49}}$ means the principal or nonnegative square root of $\frac{36}{49}$. Therefore, $\sqrt{\frac{36}{49}} = \frac{6}{7}$.

9. $\sqrt{24} = \sqrt{4}\sqrt{6} = 2\sqrt{6}$

13. $-3\sqrt{44} = -3\sqrt{4}\sqrt{11} = -3(2)\sqrt{11} = -6\sqrt{11}$

17. $\sqrt{12x^2} = \sqrt{4x^2}\sqrt{3} = 2x\sqrt{3}$

21. $\frac{3}{7}\sqrt{45xy^6} = \frac{3}{7}\sqrt{9y^6}\sqrt{5x} = \frac{3}{7}(3y^3)\sqrt{5x} = \frac{9y^3\sqrt{5x}}{7}$

25. $\sqrt[3]{16x^4} = \sqrt[3]{8x^3}\,\sqrt[3]{2x} = 2x\sqrt[3]{2x}$

29. $\sqrt{\frac{12}{25}} = \frac{\sqrt{12}}{\sqrt{25}} = \frac{\sqrt{4}\sqrt{3}}{5} = \frac{2\sqrt{3}}{5}$

33. $\frac{4\sqrt{3}}{\sqrt{5}} = \frac{4\sqrt{3}}{\sqrt{5}} \cdot \frac{\sqrt{5}}{\sqrt{5}} = \frac{4\sqrt{15}}{5}$

37. $\frac{\sqrt{5}}{\sqrt{12x^4}} = \frac{\sqrt{5}}{\sqrt{4x^4}\sqrt{3}} = \frac{\sqrt{5}}{2x^2\sqrt{3}} = \frac{\sqrt{5}}{2x^2\sqrt{3}} \cdot \frac{\sqrt{3}}{\sqrt{3}} = \frac{\sqrt{15}}{6x^2}$

<u>or</u>

$\frac{\sqrt{5}}{\sqrt{12x^4}} = \frac{\sqrt{5}}{\sqrt{12x^4}} \cdot \frac{\sqrt{3}}{\sqrt{3}} = \frac{\sqrt{15}}{\sqrt{36x^4}} = \frac{\sqrt{15}}{6x^2}$

41. $\frac{\sqrt[3]{27}}{\sqrt[3]{4}} = \frac{3}{\sqrt[3]{4}} = \frac{3}{\sqrt[3]{4}} \cdot \frac{\sqrt[3]{2}}{\sqrt[3]{2}} = \frac{3\sqrt[3]{2}}{\sqrt[3]{8}} = \frac{3\sqrt[3]{2}}{2}$

45. $5\sqrt{12} + 2\sqrt{3} = 5\sqrt{4}\sqrt{3} + 2\sqrt{3} = 10\sqrt{3} + 2\sqrt{3} = 12\sqrt{3}$

49. $\frac{5}{6}\sqrt{48} - \frac{3}{4}\sqrt{12} = \frac{5}{6}\sqrt{16}\sqrt{3} - \frac{3}{4}\sqrt{4}\sqrt{3} = \frac{5}{6}(4)\sqrt{3} - \frac{3}{4}(2)\sqrt{3}$

$= \frac{10}{3}\sqrt{3} - \frac{3}{2}\sqrt{3} = \frac{20}{6}\sqrt{3} - \frac{9}{6}\sqrt{3} = \frac{11\sqrt{3}}{6}$

53. $(4\sqrt{3})(6\sqrt{8}) = 4\cdot6\cdot\sqrt{3}\cdot\sqrt{8} = 24\sqrt{24} = 24\sqrt{4}\sqrt{6} = 48\sqrt{6}$

57. $3\sqrt{x}(\sqrt{6xy} - \sqrt{8y}) = (3\sqrt{x})(\sqrt{6xy})-(3\sqrt{x})(\sqrt{8y}) = 3\sqrt{6x^2y} - 3\sqrt{8xy} = 3x\sqrt{6y} - 6\sqrt{2xy}$

61. $(4\sqrt{2} + \sqrt{3})(3\sqrt{2} + 2\sqrt{3}) = 24 + 8\sqrt{6} + 3\sqrt{6} + 6 = 30 + 11\sqrt{6}$

65. $(\sqrt{x} + \sqrt{y})^2 = (\sqrt{x})^2 + 2\sqrt{x}\sqrt{y} + (\sqrt{y})^2 = x + 2\sqrt{xy} + y$

69. $\dfrac{3}{\sqrt{5}+2} = \dfrac{3}{\sqrt{5}+2} \cdot \dfrac{\sqrt{5}-2}{\sqrt{5}-2} = \dfrac{3\sqrt{5}-6}{5-4} = 3\sqrt{5} - 6$

73. $\dfrac{\sqrt{2}}{2\sqrt{5}+3\sqrt{7}} = \dfrac{\sqrt{2}}{2\sqrt{5}+3\sqrt{7}} \cdot \dfrac{2\sqrt{5}-3\sqrt{7}}{2\sqrt{5}-3\sqrt{7}} = \dfrac{2\sqrt{10}-3\sqrt{14}}{20-63} = \dfrac{2\sqrt{10}-3\sqrt{14}}{-43} = \dfrac{-2\sqrt{10}+3\sqrt{14}}{43}$

77. $\dfrac{\sqrt{x}}{\sqrt{x}+\sqrt{y}} = \dfrac{\sqrt{x}}{\sqrt{x}+\sqrt{y}} \cdot \dfrac{\sqrt{x}-\sqrt{y}}{\sqrt{x}-\sqrt{y}} = \dfrac{x-\sqrt{xy}}{x-y}$

81. $\left(\dfrac{\sqrt{2x+2h} - \sqrt{2x}}{h}\right)\left(\dfrac{\sqrt{2x+2h}+\sqrt{2x}}{\sqrt{2x+2h}+\sqrt{2x}}\right) = \dfrac{2x+2h-2x}{h(\sqrt{2x+2h}+\sqrt{2x})} = \dfrac{2h}{h(\sqrt{2x+2h}+\sqrt{2x})} = \dfrac{2}{\sqrt{2x+2h}+\sqrt{2x}}$

Problem Set 1.7

1. $49^{\frac{1}{2}} = \sqrt{49} = 7$

5. $-8^{\frac{2}{3}} = -\sqrt[3]{8^2} = -(\sqrt[3]{8})^2 = -(2)^2 = -4$

9. $16^{\frac{3}{2}} = (\sqrt{16})^3 = 4^3 = 64$

13. $64^{-\frac{5}{6}} = \dfrac{1}{64^{\frac{5}{6}}} = \dfrac{1}{(\sqrt[6]{64})^5} = \dfrac{1}{(2)^5} = \dfrac{1}{32}$

17. $\left(3x^{\frac{1}{4}}\right)\left(5x^{\frac{1}{3}}\right) = 15x^{\frac{1}{4}+\frac{1}{3}} = 15x^{\frac{7}{12}}$

21. $\left(4x^{\frac{1}{4}}y^{\frac{1}{2}}\right)^3 = (4)^3\left(x^{\frac{1}{4}}\right)^3\left(y^{\frac{1}{2}}\right)^3 = 64x^{\frac{3}{4}}y^{\frac{3}{2}}$

25. $\dfrac{56a^{\frac{1}{6}}}{8a^{\frac{1}{4}}} = 7a^{\frac{1}{6}-\frac{1}{4}} = 7a^{-\frac{1}{12}} = \dfrac{7}{a^{\frac{1}{12}}}$

29. $\left(\dfrac{x^2}{y^3}\right)^{-\frac{1}{2}} = \dfrac{(x^2)^{-\frac{1}{2}}}{(y^3)^{-\frac{1}{2}}} = \dfrac{x^{-1}}{y^{-\frac{3}{2}}} = \dfrac{y^{\frac{3}{2}}}{x}$

33. $\sqrt{2}\,\sqrt[4]{2} = \left(2^{\frac{1}{2}}\right)\left(2^{\frac{1}{4}}\right) = 2^{\frac{1}{2}+\frac{1}{4}} = 2^{\frac{3}{4}} = \sqrt[4]{2^3} = \sqrt[4]{8}$

37. $\sqrt{xy}\,\sqrt[4]{x^3y^5} = (xy)^{\frac{1}{2}}(x^3y^5)^{\frac{1}{4}} = x^{\frac{1}{2}}y^{\frac{1}{2}}x^{\frac{3}{4}}y^{\frac{5}{4}} = x^{\frac{5}{4}}y^{\frac{7}{4}} = xyx^{\frac{1}{4}}y^{\frac{3}{4}} = xy\sqrt[4]{xy^3}$

41. $\sqrt[3]{4}\sqrt{8} = (4)^{\frac{1}{3}}(8)^{\frac{1}{2}} = (2^2)^{\frac{1}{3}}(2^3)^{\frac{1}{2}} = (2^{\frac{2}{3}})(2^{\frac{3}{2}}) = 2^{\frac{13}{6}} = (2^2)(2^{\frac{1}{6}}) = 4\sqrt[6]{2}$

45. $\dfrac{\sqrt[3]{8}}{\sqrt[4]{4}} = \dfrac{(8)^{\frac{1}{3}}}{(4)^{\frac{1}{4}}} = \dfrac{(2^3)^{\frac{1}{3}}}{(2^2)^{\frac{1}{4}}} = \dfrac{2^1}{2^{\frac{1}{2}}} = 2^{\frac{1}{2}} = \sqrt{2}$

49. $\dfrac{5}{\sqrt[3]{x}} = \dfrac{5}{\sqrt[3]{x}} \cdot \dfrac{\sqrt[3]{x^2}}{\sqrt[3]{x^2}} = \dfrac{5\sqrt[3]{x^2}}{x}$

53. $\dfrac{\sqrt[4]{x^3}}{\sqrt[5]{y^3}} = \dfrac{\sqrt[4]{x^3}}{\sqrt[5]{y^3}} \cdot \dfrac{\sqrt[5]{y^2}}{\sqrt[5]{y^2}} = \dfrac{x^{\frac{3}{4}} \cdot y^{\frac{2}{5}}}{y} = \dfrac{x^{\frac{15}{20}} \cdot y^{\frac{8}{20}}}{y} = \dfrac{\sqrt[20]{x^{15}y^8}}{y}$

57. (a) $\sqrt{\sqrt[3]{2}} = \left(2^{\frac{1}{3}}\right)^{\frac{1}{2}} = 2^{\frac{1}{6}} = \sqrt[6]{2}$

65. $$\frac{(x^2+2x)^{\frac{1}{2}}-x(x+1)(x^2+2x)^{-\frac{1}{2}}}{\left[(x^2+2x)^{\frac{1}{2}}\right]^2} \cdot \frac{(x^2+2x)^{\frac{1}{2}}}{(x^2+2x)^{\frac{1}{2}}}$$

$$= \frac{(x^2+2x)^1-x(x+1)(x^2+2x)^0}{(x^2+2x)^{\frac{3}{2}}}$$ [Remember that $(x^2+2x)^0 = 1$.]

$$= \frac{x^2+2x-x^2-x}{(x^2+2x)^{\frac{3}{2}}} = \frac{x}{(x^2+2x)^{\frac{3}{2}}}$$

Problem Set 1.8

1. $(5+2i)+(8+6i) = (5+8)+(2+6)i = 13+8i$

5. $(-7-3i)+(-4+4i) = (-7+(-4)+(-3+4)i = -11+i$

9. $(-\frac{3}{4} - \frac{1}{4}i)+(\frac{3}{5} + \frac{2}{3}i) = (-\frac{3}{4} + \frac{3}{5})+(-\frac{1}{4} + \frac{2}{3})i = -\frac{3}{20} + \frac{5}{12}i$

13. $(5+3i)+(7-2i)+(-8-i) = (5+7+(-8))+(3+(-2)+(-1))i = 4+0i$

17. $\sqrt{-19} = i\sqrt{19}$

21. $\sqrt{-8} = i\sqrt{8} = i\sqrt{4}\sqrt{2} = 2i\sqrt{2}$

25. $\sqrt{-54} = i\sqrt{54} = i\sqrt{9}\sqrt{6} = 3i\sqrt{6}$

29. $4\sqrt{-18} = 4i\sqrt{18} = 4i\sqrt{9}\sqrt{2} = 12i\sqrt{2}$

33. $\sqrt{-2}\sqrt{-3} = (i\sqrt{2})(i\sqrt{3}) = i^2\sqrt{6} = -\sqrt{6}$

37. $\sqrt{-6}\sqrt{-10} = (i\sqrt{6})(i\sqrt{10}) = i^2\sqrt{60} = -\sqrt{4}\sqrt{15} = -2\sqrt{15}$

41. $\dfrac{\sqrt{-36}}{\sqrt{-4}} = \dfrac{i\sqrt{36}}{i\sqrt{4}} = \dfrac{6i}{2i} = 3$

45. $(3i)(7i) = 21i^2 = 21(-1) = -21$

49. $(3+2i)(4+6i) = 12+18i+8i+12i^2 = 12+26i+12(-1) = 0+26i$

53. $(-2-3i)(4+6i) = -8-12i-12i-18i^2 = -8-24i-18(-1) = 10-24i$

57. $(3+4i)^2 = 3^2+2(3)(4i)+(4i)^2 = 9+24i+16i^2 = 9+24i+16(-1) = -7+24i$

61. $(8-7i)(8+7i) = 8^2-(7i)^2 = 64-49i^2 = 64-49(-1) = 113+0i$

65. $\dfrac{4i}{3-2i} = \dfrac{4i}{3-2i} \cdot \dfrac{3+2i}{3+2i} = \dfrac{12i+8i^2}{9-4i^2} = \dfrac{-8+12i}{13} = -\dfrac{8}{13} + \dfrac{12}{13}i$

69. $\dfrac{3}{2i} = \dfrac{3}{2i} \cdot \dfrac{i}{i} = \dfrac{3i}{2i^2} = \dfrac{3i}{-2} = 0 - \dfrac{3}{2}i$

73. $\dfrac{4+7i}{2-3i} \cdot \dfrac{2+3i}{2+3i} = \dfrac{8+12i+14i+21i^2}{4-9i^2} = \dfrac{-13+26i}{13} = -1+2i$

77. $\dfrac{-1-i}{-2-3i} \cdot \dfrac{-2+3i}{-2+3i} = \dfrac{2-3i+2i-3i^2}{4-9i^2} = \dfrac{5-i}{13} = \dfrac{5}{13} - \dfrac{1}{13}i$

CHAPTER 2

Problem Set 2.1

1. $9x-3 = -21$
   $9x = -18$
   $x = -2$

   The solution set is $\{-2\}$.

5. $3n-2 = 2n+5$
   $n-2 = 5$
   $n = 7$

   The solution set is $\{7\}$.

9. $-3(x+1) = 7$
   $-3x-3 = 7$
   $-3x = 10$
   $x = -\frac{10}{3}$

   The solution set is $\{-\frac{10}{3}\}$.

13. $4(n-2)-3(n-1) = 2(n+6)$
    $4n-8-3n+3 = 2n+12$
    $n-5 = 2n+12$
    $-17 = n$

    The solution set is $\{-17\}$.

17. $-2(y-4)-(3y-1) = -2+5(y+1)$
    $-2y+8-3y+1 = -2+5y+5$
    $-5y+9 = 5y+3$
    $6 = 10y$
    $\frac{6}{10} = y$
    $\frac{3}{5} = y$

    The solution set is $\{\frac{3}{5}\}$.

21. $\frac{3n}{4} - \frac{n}{12} = 6$
    $12(\frac{3n}{4} - \frac{n}{12}) = 12(6)$
    $9n-n = 72$
    $8n = 72$
    $n = 9$

    The solution set is $\{9\}$.

25. $\frac{y}{5} - 2 = \frac{y}{2} + 1$
    $10(\frac{y}{5} - 2) = 10(\frac{y}{2} + 1)$
    $2y-20 = 5y+10$
    $-30 = 3y$
    $-10 = y$

    The solution set is $\{-10\}$.

29. $\frac{n-3}{2} - \frac{4n-1}{6} = \frac{2}{3}$
    $6(\frac{n-3}{2} - \frac{4n-1}{6}) = 6(\frac{2}{3})$
    $3(n-3)-(4n-1) = 4$
    $3n-9-4n+1 = 4$
    $-n-8 = 4$
    $-n = 12$
    $n = -12$

    The solution set is $\{-12\}$.

33. $\frac{3n-1}{8} - 2 = \frac{2n+5}{7}$
    $56(\frac{3n-1}{8} - 2) = 56(\frac{2n+5}{7})$
    $7(3n-1)-112 = 8(2n+5)$
    $21n-7-112 = 16n+40$
    $21n-119 = 16n+40$
    $5n = 159$
    $n = \frac{159}{5}$

    The solution set is $\{\frac{159}{5}\}$.

37. $\frac{2x+1}{14} - \frac{3x+4}{7} = \frac{x-1}{2}$
    $14(\frac{2x+1}{14} - \frac{3x+4}{7}) = 14(\frac{x-1}{2})$
    $2x+1-2(3x+4) = 7(x-1)$
    $2x+1-6x-8 = 7x-7$
    $-4x-7 = 7x-7$
    $0 = 11x$
    $0 = x$

    The solution set is $\{0\}$.

41.
$$\begin{aligned}(2y+1)(3y-2)-(6y-1)(y+4) &= -20y\\ 6y^2-y-2-6y^2-23y+4 &= -20y\\ -24y+2 &= -20y\\ 2 &= 4y\\ \frac{2}{4} &= y\\ \frac{1}{2} &= y\end{aligned}$$

The solution set is $\{\frac{1}{2}\}$.

45. Let n represent the smaller number. Then 4n-5 represents the larger number.

$$\begin{aligned}n+(4n-5) &= 30\\ n+4n-5 &= 30\\ 5n &= 35\\ n &= 7\end{aligned}$$

The smaller number is 7 and the larger is 4(7)-5 = 23.

49. Let n represent the smallest integer. Then n+2 and n+4 represent the other two.

$$\begin{aligned}3(n+4) &= 2(n+n+2)-23\\ 3n+12 &= 2(2n+2)-23\\ 3n+12 &= 4n+4-23\\ 3n+12 &= 4n-19\\ 31 &= n\end{aligned}$$

The integers are 31, 33, and 35.

53. Let x represent Renee's salary. Then x-4000 represents Kelly's salary and $\frac{2}{3}x$ represents Nina's salary.

$$\begin{aligned}\frac{x+(x-4000)+\frac{2}{3}x}{3} &= 20000\\ x+x-4000-\frac{2}{3}x &= 60000\\ 2\frac{2}{3}x &= 64000\\ x &= 24000\end{aligned}$$

Renee's salary is \$24,000, Kelly's salary is \$24,000-\$4000 = \$20,000, and Nina's salary is $\frac{2}{3}$(\$24,000) = \$16,000.

57. Let f represent the number of females. Then 2f-8 represents the number of males.

$$\begin{aligned}f+(2f-8) &= 43\\ 3f-8 &= 43\\ 3f &= 51\\ f &= 17\end{aligned}$$

There are 17 females and 2(17)-8 = 26 males.

61. Let x represent Pedro's present age. Then x+6 represents Brad's present age. Also, x-5 represents Pedro's age 5 years ago, and x+6-5 represents Brad's age 5 years ago.

$$\begin{aligned}x-5 &= \frac{3}{4}(x+1) \qquad [x+6-5 = x+1]\\ 4x-20 &= 3x+3\\ x &= 23\end{aligned}$$

Pedro's present age is 23 and Brad's present age is 23+6 = 29.

Problem Set 2.2

1. $$\frac{x-2}{3} + \frac{x+1}{4} = \frac{1}{6}$$

$$12(\frac{x-2}{3} + \frac{x+1}{4}) = 12(\frac{1}{6})$$

$$4(x-2)+3(x+1) = 2$$
$$4x-8+3x+3 = 2$$
$$7x-5 = 2$$
$$7x = 7$$
$$x = 1$$

The solution set is {1}.

5. $$\frac{1}{3n} + \frac{1}{2n} = \frac{1}{4}, \; n \neq 0.$$

$$12n(\frac{1}{3n} + \frac{1}{2n}) = 12n(\frac{1}{4})$$

$$4+6 = 3n$$
$$10 = 3n$$
$$\frac{10}{3} = n$$

The solution set is $\{\frac{10}{3}\}$.

9. $$\frac{n+67}{n} = 5 + \frac{11}{n}, \; n \neq 0$$

$$n(\frac{n+67}{n}) = n(5 + \frac{11}{n})$$

$$n+67 = 5n+11$$
$$56 = 4n$$
$$14 = n$$

The solution set is {14}.

13. $$\frac{4}{2y-3} - \frac{7}{3y-5} = 0, \; y \neq \frac{3}{2} \text{ and } y \neq \frac{5}{3}$$

$$\frac{4}{2y-3} = \frac{7}{3y-5}$$

$$4(3y-5) = 7(2y-3)$$
$$12y-20 = 14y-21$$
$$1 = 2y$$
$$\frac{1}{2} = y$$

The solution set is $\{\frac{1}{2}\}$.

17. $$\frac{3x}{2x-1} - 4 = \frac{x}{2x-1}, \; x \neq \frac{1}{2}$$

$$(2x-1)(\frac{3x}{2x-1} - 4) = (2x-1)(\frac{x}{2x-1})$$

$$3x-4(2x-1) = x$$
$$3x-8x+4 = x$$
$$-5x+4 = x$$
$$4 = 6x$$
$$\frac{4}{6} = x$$
$$\frac{2}{3} = x$$

The solution set is $\{\frac{2}{3}\}$.

21. $$\frac{n}{n-3} - \frac{3}{2} = \frac{3}{n-3}, \; n \neq 3$$

$$2(n-3)(\frac{n}{n-3} - \frac{3}{2}) = 2(n-3)(\frac{3}{n-3})$$

$$2n-3(n-3) = 6$$
$$2n-3n+9 = 6$$
$$-n = -3$$
$$n = 3$$

Since the initial restriction was $n \neq 3$, (3 makes a denominator of 0) the solution set is $\emptyset$.

25. $$.09x+.1(700-x) = 67$$
$$100[.09x+.1(700-x)] = 100(67)$$
$$9x+10(700-x) = 6700$$
$$9x+7000-10x = 6700$$
$$-x = -300$$
$$x = 300$$

The solution set is {300}.

29. $$.8(t-2) = .5(9t+10)$$
$$10[.8(t-2)] = 10[.5(9t+10)]$$
$$8(t-2) = 5(9t+10)$$
$$8t-16 = 45t+50$$
$$-66 = 37t$$
$$-\frac{66}{37} = t$$

The solution set is $\{-\frac{66}{37}\}$.

33.

$$P = 2\ell+2w$$
$$P-2\ell = 2w$$
$$\frac{P-2\ell}{2} = w$$

37.

$$C = \frac{5}{9}(F-32)$$
$$9C = 5F-160$$
$$9C+160 = 5F$$
$$\frac{9C+160}{5} = F$$

41.

$$I = k\ell(T-t)$$
$$I = k\ell T-k\ell t$$
$$I+k\ell t = k\ell T$$
$$\frac{I+k\ell t}{k\ell} = T$$

45. Let n represent the smaller number. Then 98-n represents the larger number.

$$\frac{98-n}{n} = 4 + \frac{13}{n}, \quad n \neq 0$$
$$98-n = 4n+13$$
$$85 = 5n$$
$$17 = n$$

The numbers are 17 and 98-17 = 81.

49. Let x represent part of the money. Then 2250-x represents the other part.

$$\frac{x}{2250-x} = \frac{2}{3}$$
$$3x = 2(2250-x)$$
$$3x = 4500-2x$$
$$5x = 4500$$
$$x = 900$$

One person receives \$900 and the other person receives 2250-900 = \$1350.

53. Let p represent the original price. A 20% discount means that he paid 80% of the original price.

$$.8p = 52$$
$$p = 65$$

The original price was \$65.

57. Let s represent the selling price. We can use the relationship "selling price equals cost plus profit", where the profit is 15% of the cost, as a guideline.

$$s = 28+.15(28)$$
$$s = 28+4.2$$
$$s = 32.2$$

The selling price should be \$32.20.

61. Let s represent the selling price. In this problem, the profit is stated as a percent of the selling price.

$$s = 18+.4s$$
$$.6s = 18$$
$$s = 30$$

The selling price should be \$30.

65. Let d represent the number of dimes. Then 3d represents the number of quarters and 70-4d represents the number of half dollars.

$$10d+25(3d)+50(70-4d) = 1775$$
$$10d+75d+3500-200d = 1775$$
$$-115d = -1725$$
$$d = 15$$

There are 15 dimes, 3(15) = 45 quarters, and 70-4(15) = 10 half dollars.

69. Let x represent the amount to be invested at 12%.

$$\begin{aligned} .11(2500)+.12x &= 695 \\ 275+.12x &= 695 \\ .12x &= 420 \\ x &= 3500 \end{aligned}$$

She should invest $3500 at 12%.

Problem Set 2.3

1.
$$\begin{aligned} x^2-3x-28 &= 0 \\ (x-7)(x+4) &= 0 \\ x-7 = 0 \text{ or } x+4 &= 0 \\ x = 7 \text{ or } x &= -4 \end{aligned}$$

The solution set is $\{-4,7\}$.

5.
$$\begin{aligned} 2x^2-3x &= 0 \\ x(2x-3) &= 0 \\ x = 0 \text{ or } 2x-3 &= 0 \\ x = 0 \text{ or } 2x &= 3 \\ x = 0 \text{ or } x &= \frac{3}{2} \end{aligned}$$

The solution set is $\{0,\frac{3}{2}\}$.

9.
$$\begin{aligned} (2n+1)^2 &= 20 \\ (2n+1) &= \pm\sqrt{20} = \pm 2\sqrt{5} \\ 2n+1 = -2\sqrt{5} &\text{ or } 2n+1 = 2\sqrt{5} \\ 2n = -1-2\sqrt{5} &\text{ or } 2n = -1+2\sqrt{5} \\ n = \frac{-1-2\sqrt{5}}{2} &\text{ or } n = \frac{-1+2\sqrt{5}}{2} \end{aligned}$$

The solution set is $\{\frac{-1\pm 2\sqrt{5}}{2}\}$.

13.
$$\begin{aligned} (x-2)^2 &= -4 \\ x-2 &= \pm\sqrt{-4} = \pm 2i \\ x-2 = -2i &\text{ or } x-2 = 2i \\ x = 2-2i &\text{ or } x = 2+2i \end{aligned}$$

The solution set is $\{2\pm 2i\}$.

17.
$$\begin{aligned} x^2-10x+24 &= 0 \\ x^2-10x+25 &= -24+25 \\ (x-5)^2 &= 1 \\ x-5 &= \pm 1 \\ x-5 = -1 \text{ or } x-5 &= 1 \\ x = 4 \text{ or } x &= 6 \end{aligned}$$

The solution set is $\{4,6\}$.

21.
$$\begin{aligned} y^2-3y &= -1 \\ y^2-3y+\frac{9}{4} &= -1+\frac{9}{4} \\ (y-\frac{3}{2})^2 &= \frac{5}{4} \\ y-\frac{3}{2} &= \pm\frac{\sqrt{5}}{2} \\ y-\frac{3}{2} = -\frac{\sqrt{5}}{2} &\text{ or } y-\frac{3}{2} = \frac{\sqrt{5}}{2} \\ y = \frac{3}{2}-\frac{\sqrt{5}}{2} &\text{ or } y = \frac{3}{2}+\frac{\sqrt{5}}{2} \\ y = \frac{3-\sqrt{5}}{2} &\text{ or } y = \frac{3+\sqrt{5}}{2} \end{aligned}$$

The solution set is $\{\frac{3\pm\sqrt{5}}{2}\}$.

25.

$$2t^2+12t-5 = 0$$
$$t^2+6t = \frac{5}{2}$$
$$t^2+6t+9 = \frac{5}{2} + 9$$
$$(t+3)^2 = \frac{23}{2}$$
$$t+3 = \frac{\pm\sqrt{46}}{2}$$
$$t+3 = -\frac{\sqrt{46}}{2} \text{ or } t+3 = \frac{\sqrt{46}}{2}$$
$$t = -3 - \frac{\sqrt{46}}{2} \text{ or } t = -3 + \frac{\sqrt{46}}{2}$$
$$t = \frac{-6-\sqrt{46}}{2} \text{ or } t = \frac{-6+\sqrt{46}}{2}$$

The solution set is $\{\frac{-6 \pm \sqrt{46}}{2}\}$.

29.

$$3n^2+5n-1 = 0$$
$$n^2+\frac{5}{3}n = \frac{1}{3}$$
$$n^2+\frac{5}{3}n+\frac{25}{36} = \frac{1}{3}+\frac{25}{36}$$
$$(n+\frac{5}{6})^2 = \frac{37}{36}$$
$$n+\frac{5}{6} = \pm\frac{\sqrt{37}}{6}$$
$$n+\frac{5}{6} = -\frac{\sqrt{37}}{6} \text{ or } n+\frac{5}{6} = \frac{\sqrt{37}}{6}$$
$$n = \frac{-5-\sqrt{37}}{6} \text{ or } n = \frac{-5+\sqrt{37}}{6}$$

The solution set is $\{\frac{-5 \pm \sqrt{37}}{6}\}$.

33.

$$3x^2+16x = -5$$
$$3x^2+16x+5 = 0$$
$$x = \frac{-16 \pm \sqrt{256-60}}{6} = \frac{-16 \pm \sqrt{196}}{6} = \frac{-16 \pm 14}{6}$$
$$x = \frac{-16-14}{6} = -5 \text{ or } x = \frac{-16+14}{6} = -\frac{1}{3}$$

The solution set is $\{-5, -\frac{1}{3}\}$.

37.

$$2a^2-6a+1 = 0$$
$$a = \frac{-(-6) \pm \sqrt{36-8}}{4} = \frac{6 \pm \sqrt{28}}{4} = \frac{6 \pm 2\sqrt{7}}{4} = \frac{3 \pm \sqrt{7}}{2}$$

The solution set is $\{\frac{3 \pm \sqrt{7}}{2}\}$.

41.

$$x^2+4 = 8x$$
$$x^2-8x+4 = 0$$
$$x = \frac{-(-8) \pm \sqrt{64-16}}{2} = \frac{8 \pm \sqrt{48}}{2} = \frac{8 \pm 4\sqrt{3}}{2} = 4 \pm 2\sqrt{3}$$

The solution set is $\{4 \pm 2\sqrt{3}\}$.

45.

$$8x^2+10x-3 = 0$$
$$(4x-1)(2x+3) = 0$$
$$4x-1 = 0 \text{ or } 2x+3 = 0$$
$$4x = 1 \text{ or } 2x = -3$$
$$x = \frac{1}{4} \text{ or } x = -\frac{3}{2}$$

The solution set is $\{-\frac{3}{2}, \frac{1}{4}\}$.

49. $2t^2-3t+7 = 0$

$$t = \frac{-(-3) \pm \sqrt{9-56}}{4} = \frac{3 \pm \sqrt{-47}}{4} = \frac{3 \pm i\sqrt{47}}{4}$$

The solution set is $\{\frac{3 \pm i\sqrt{47}}{4}\}$.

53. $4y^2+4y-1 = 0$

$$y = \frac{-4 \pm \sqrt{16+16}}{8} = \frac{-4 \pm \sqrt{32}}{8} = \frac{-4 \pm 4\sqrt{2}}{8} = \frac{-1 \pm \sqrt{2}}{2}$$

The solution set is $\{\frac{-1 \pm \sqrt{2}}{2}\}$.

57.

$$t^2+20t = 25$$

$$t^2+20t+100 = 25+100$$

$$(t+10)^2 = 125$$

$$t+10 = \pm\sqrt{125} = \pm 5\sqrt{5}$$

$$t = -10 \pm 5\sqrt{5}$$

The solution set is $\{-10 \pm 5\sqrt{5}\}$.

61. (a) $4x^2+20x+25 = 0$

$$b^2-4ac = 20^2-4(4)(25) = 400-400 = 0$$

Therefore, the equation has one real solution.

65. Let n represent one of the numbers. Then 22-n represents the other number.

$$n(22-n) = 112$$

$$-n^2+22n = 112$$

$$n^2-22n = -112$$

$$n^2-22n+121 = -112+121$$

$$(n-11)^2 = 9$$

$$n-11 = \pm 3$$

$n-11 = -3$ or $n-11 = 3$

$n = 8$ or $n = 14$

If $n = 8$, then $22-8 = 14$.

If $n = 14$, then $22-14 = 8$.

The numbers are 8 and 14.

69. Let w represent the width of the rectangle. Then 22-w represents the length. (Remember that length plus width equals one-half of the perimeter.)

$$w(22-w) = 112$$

$$-w^2+22w = 112$$

$$w^2-22w = -112$$

$$w^2-22w+121 = -112+121$$

$$(w-11)^2 = 9$$

$$w-11 = \pm 3$$

$$w = 11 \pm 3$$

$w = 14$ or $w = 8$

The rectangle is 8 inches by 14 inches.

73. A diagram may help with such a problem.

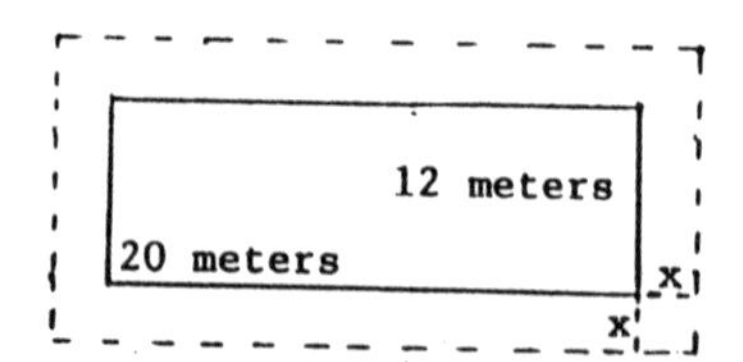

The area of the sidewalk equals the area of the large rectangle minus the area of the small rectangle.

$$(20+2x)(12+2x)-20(12) = 68$$
$$240+64x+4x^2-240 = 68$$
$$4x^2+64x-68 = 0$$
$$x^2+16x-17 = 0$$
$$(x+17)(x-1) = 0$$
$$x+17 = 0 \text{ or } x-1 = 0$$
$$x = -17 \text{ or } x = 1$$

Disregarding the negative answer, the width of the sidewalk must be 1 meter.

77. Let r represent the length of a radius. Then the statement "area equals four times circumference" translates into

$$\pi r^2 = 4(2\pi r).$$

Now we can solve this equation for r.

$$\pi r^2 = 8\pi r$$
$$\pi r^2-8\pi r = 0$$
$$\pi r(r-8) = 0$$
$$\pi r = 0 \text{ or } r-8 = 0$$
$$r = 0 \text{ or } r = 8$$

The length of a radius is 8 units.

Problem Set 2.4

1.
$$\frac{x}{2x-8} + \frac{16}{x^2-16} = \frac{1}{2}$$
$$\frac{x}{2(x-4)} + \frac{16}{(x+4)(x-4)} = \frac{1}{2}, \quad x \neq 4 \text{ and } x \neq -4$$
$$2(x-4)(x+4)\left[\frac{x}{2(x-4)} + \frac{16}{(x+4)(x-4)}\right] = 2(x-4)(x+4)\left(\frac{1}{2}\right)$$
$$x(x+4)+2(16) = (x-4)(x+4)$$
$$x^2+4x+32 = x^2-16$$
$$4x = -48$$
$$x = -12$$

The solution set is $\{-12\}$.

5. $$2 + \frac{4}{n-2} = \frac{8}{n(n-2)}, \quad n \neq 2 \text{ and } n \neq 0$$

$$n(n-2)[2 + \frac{4}{n-2}] = n(n-2)[\frac{8}{n(n-2)}]$$

$$2n(n-2)+4n = 8$$

$$2n^2-4n+4n = 8$$

$$2n^2 = 8$$

$$n^2 = 4$$

$$n = \pm 2$$

The initial restriction states that $n \neq 2$; therefore, the solution set is $\{-2\}$.

9. $$\frac{-2}{3x+2} + \frac{x-1}{(3x+2)(3x-2)} = \frac{3}{4(3x-2)}, \quad x \neq -\frac{2}{3} \text{ and } x \neq \frac{2}{3}$$

$$4(3x+2)(3x-2)[\frac{-2}{3x+2} + \frac{x-1}{(3x+2)(3x-2)}] = 4(3x+2)(3x-2)[\frac{3}{4(3x-2)}]$$

$$-8(3x-2)+4(x-1) = 3(3x+2)$$

$$-24x+16+4x-4 = 9x+6$$

$$-20x+12 = 9x+6$$

$$6 = 29x$$

$$\frac{6}{29} = x$$

The solution set is $\{\frac{6}{29}\}$.

13. $$\frac{3y+1}{(3y+2)(y-2)} + \frac{9}{(3y+2)(3y-2)} = \frac{2y-2}{(3y-2)(y-2)}, \quad y \neq -\frac{2}{3}, \text{ and } y \neq 2 \text{ and } y \neq \frac{2}{3}$$

$$(3y+2)(y-2)(3y-2)[\frac{3y+1}{(3y+2)(y-2)} + \frac{9}{(3y+2)(3y-2)}] =$$

$$(3y+2)(y-2)(3y-2)[\frac{2y-2}{(3y-2)(y-2)}]$$

$$(3y-2)(3y+1)+9(y-2) = (3y+2)(2y-2)$$

$$9y^2-3y-2+9y-18 = 6y^2-2y-4$$

$$9y^2+6y-20 = 6y^2-2y-4$$

$$3y^2+8y-16 = 0$$

$$(3y-4)(y+4) = 0$$

$$3y-4 = 0 \text{ or } y+4 = 0$$

$$3y = 4 \text{ or } y = -4$$

$$y = \frac{4}{3} \text{ or } y = -4$$

The solution set is $\{-4, \frac{4}{3}\}$.

17. $$\frac{7x+2}{(4x-3)(3x+5)} - \frac{1}{3x+5} = \frac{2}{4x-3}, \quad x \neq \frac{3}{4} \text{ and } x \neq -\frac{5}{3}.$$

$$(4x-3)(3x+5)\left[\frac{7x+2}{(4x-3)(3x+5)} - \frac{1}{3x+5}\right] = (4x-3)(3x+5)\left[\frac{2}{4x-3}\right]$$

$$\begin{aligned} 7x+2-1(4x-3) &= 2(3x+5) \\ 7x+2-4x+3 &= 6x+10 \\ 3x+5 &= 6x+10 \\ -5 &= 3x \\ -\frac{5}{3} &= x \end{aligned}$$

Since the initial restriction stated that $x \neq -\frac{5}{3}$, the solution set is $\emptyset$.

21. Let r represent the number of rows. Then 2r-4 represents the number of trees per row. The total number of trees (126) equals the number of rows times the number of trees per row.

$$\begin{aligned} r(2r-4) &= 126 \\ 2r^2-4r-126 &= 0 \\ r^2-2r-63 &= 0 \\ (r-9)(r+7) &= 0 \\ r-9 = 0 \text{ or } r+7 &= 0 \\ r = 9 \text{ or } r &= -7 \end{aligned}$$

The negative solution must be disregarded. Therefore, there are 9 rows and $2(9)-4 = 14$ trees per row.

25. Let t represent Rita's time bicycling out into the country. Then $5\frac{5}{6} - t$ represents her time returning.

	distance	rate	time
out	$20t$	20	t
back	$15(\frac{35}{6} - t)$	15	$\frac{35}{6} - t$

Since the distance "out" equals the distance "back", we can equate the two distances.

$$\begin{aligned} 20t &= 15(\frac{35}{6} - t) \\ 6(20t) &= 6[15(\frac{35}{6} - t)] \\ 120t &= 90(\frac{35}{6} - t) \\ 120t &= 525-90t \\ 210t &= 525 \\ t &= \frac{525}{210} = \frac{5}{2} \end{aligned}$$

If $t = \frac{5}{2}$, then $20t = 20(\frac{5}{2}) = 50$. Thus, the distance she rode out into the country was 50 miles.

29. Let x represent the amount of pure alcohol to be added.

$$\begin{pmatrix}\text{pure alcohol}\\ \text{to start with}\end{pmatrix} + \begin{pmatrix}\text{pure alcohol}\\ \text{to be added}\end{pmatrix} = \begin{pmatrix}\text{pure alcohol in}\\ \text{final solution}\end{pmatrix}$$

$$(40\%)(6) \quad + \quad x \quad = \quad 60\%(6+x)$$

Solving this equation produces

$$\begin{aligned} .4(6)+x &= .6(6+x) \\ 10[.4(6)+x] &= 10[.6(6+x)] \\ 24+10x &= 36+6x \\ 4x &= 12 \\ x &= 3 \end{aligned}$$

We must add 3 liters of pure alcohol.

33. Let x represent the amount of solution to be drained out and replaced with pure antifreeze.

$$\begin{aligned} 40\%(10)-40\%(x)+x &= 70\%(10) \\ .4(10)-.4x+x &= .7(10) \\ 4+.6x &= 7 \\ .6x &= 3 \\ x &= 5 \end{aligned}$$

We must drain out 5 quarts of the mixture and add 5 quarts of pure antifreeze.

37. Let m represent the number of minutes before the tank overflows.

$$\frac{m}{10} - \frac{m}{12} = 1$$

$$60\left(\frac{m}{10} - \frac{m}{12}\right) = 60(1)$$

$$\begin{aligned} 6m-5m &= 60 \\ m &= 60 \end{aligned}$$

It will take 60 minutes.

41. Let t represent the time it takes Paul to type 600 words. Then t-5 represents the time it takes Amelia to type 600 words.

	quantity	rate	time
Amelia	600	$\frac{600}{t-5}$	t-5
Paul	600	$\frac{600}{t}$	t

Since Amelia's rate is 20 words per minute more than Paul's rate, we can set up and solve the following equation.

$$\frac{600}{t-5} = \frac{600}{t} + 20$$

$$t(t-5)\left[\frac{600}{t-5}\right] = t(t-5)\left[\frac{600}{t} + 20\right]$$

$$600t = 600(t-5)+20t(t-5)$$

$$600t = 600t-3000+20t^2-100t$$

$$0 = 20t^2-100t-3000$$

$$0 = t^2-5t-150$$
$$0 = (t-15)(t+10)$$
$$t-15 = 0 \text{ or } t+10 = 0$$
$$t = 15 \text{ or } t = -10$$

Disregarding the negative answer, we have $t = 15$. Therefore, Amelia's rate is $\frac{600}{15-5} = \frac{600}{10} = 60$ words per minute and Paul's rate is $\frac{600}{15} = 40$ words per minute.

45. Let h represent the number of hours that Todd had predicted it would take him. Then $h+4$ represents the time it actually took him. Also, $\frac{480}{h}$ represents his predicted hourly rate and $\frac{480}{h+4}$ represents his actual hourly rate.

$$\frac{480}{h+4} = \frac{480}{h} - \frac{1}{2} \qquad (\$.50 \text{ is } \tfrac{1}{2} \text{ of a dollar.})$$

$$2h(h+4)\left[\frac{480}{h+4}\right] = 2h(h+4)\left[\frac{480}{h} - \frac{1}{2}\right]$$

$$960h = 960(h+4)-h(h+4)$$

$$960h = 960h+3840-h^2-4h$$

$$h^2+4h-3840 = 0$$
$$(h+64)(h-60) = 0$$
$$h+64 = 0 \text{ or } h-60 = 0$$
$$h = -64 \text{ or } h = 60$$

Disregarding the negative answer, we see that Todd had predicted that it would take him 60 hours.

Problem Set 2.5

1.

$$x^3+x^2-4x-4 = 0$$

$$x^2(x+1)-4(x+1) = 0$$

$$(x+1)(x^2-4) = 0$$
$$(x+1)(x+2)(x-2) = 0$$
$$x+1 = 0 \text{ or } x+2 = 0 \text{ or } x-2 = 0$$
$$x = -1 \text{ or } x = -2 \text{ or } x = 2$$

The solution set is $\{-2,-1,2\}$.

5.
$$8x^5+10x^4 = 4x^3+5x^2$$
$$8x^5+10x^4-4x^3-5x^2 = 0$$
$$x^2(8x^3+10x^2-4x-5) = 0$$
$$x^2[2x^2(4x+5)-1(4x-5)] = 0$$
$$x^2[(2x^2-1)(4x+5)] = 0$$
$$x^2 = 0 \text{ or } 2x^2-1 = 0 \text{ or } 4x+5 = 0$$
$$x = 0 \text{ or } 2x^2 = 1 \text{ or } 4x = -5$$
$$x = 0 \text{ or } x = \frac{\pm\sqrt{2}}{2} \text{ or } x = -\frac{5}{4}$$

The solution set is $\{-\frac{5}{4}, 0, \pm\frac{\sqrt{2}}{2}\}$.

9.
$$n^{-2} = n^{-3}$$
$$\frac{1}{n^2} = \frac{1}{n^3}, \quad n \neq 0$$
$$n^3 = n^2$$
$$n^3-n^2 = 0$$
$$n^2(n-1) = 0$$
$$n^2 = 0 \text{ or } n-1 = 0$$
$$n = 0 \text{ or } n = 1$$

The initial restriction of $n \neq 0$ eliminates 0; thus, the solution set is $\{1\}$.

13.
$$\sqrt{3x-8} - \sqrt{x-2} = 0$$
$$\sqrt{3x-8} = \sqrt{x-2}$$
$$3x-8 = x-2 \quad \text{Square both sides.}$$
$$2x = 6$$
$$x = 3$$

Check: $\sqrt{9-8} - \sqrt{3-2} \stackrel{?}{=} 0$
$$1 - 1 = 0$$

The solution set is $\{3\}$.

17.
$$\sqrt[3]{2x+3} + 3 = 0$$
$$\sqrt[3]{2x+3} = -3$$
$$2x+3 = -27 \quad \text{Cube both sides.}$$
$$2x = -30$$
$$x = -15$$

Check: $\sqrt[3]{-30+3} + 3 \stackrel{?}{=} 0$
$$\sqrt[3]{-27} \stackrel{?}{=} 0$$
$$-3+3 = 0$$

The solution set is $\{-15\}$.

21.
$$\sqrt{3x-2} = 3x-2$$
$$3x-2 = 9x^2-12x+4 \quad \text{Square both sides.}$$
$$0 = 9x^2-15x+6$$
$$0 = 3x^2-5x+2$$
$$0 = (3x-2)(x-1)$$
$$3x-2 = 0 \text{ or } x-1 = 0$$
$$3x = 2 \text{ or } x = 1$$
$$x = \frac{2}{3} \text{ or } x = 1$$

Check: $\sqrt{3(\frac{2}{3})-2} \stackrel{?}{=} 3(\frac{2}{3})-2$
$$\sqrt{2-2} \stackrel{?}{=} 2-2$$
$$\sqrt{0} = 0$$
$$\sqrt{3(1)-2} \stackrel{?}{=} 3(1)-2$$
$$\sqrt{3-2} \stackrel{?}{=} 3-2$$
$$\sqrt{1} = 1$$

The solution set is $\{\frac{2}{3}, 1\}$.

25. $\sqrt{x+2} - 1 = \sqrt{x-3}$

$x+2-2\sqrt{x+2} + 1 = x-3$ Square both sides.

$x+3-2\sqrt{x+2} = x-3$

$-2\sqrt{x+2} = -6$

$\sqrt{x+2} = 3$

$x+2 = 9$ Square both sides again.

$x = 7$

Check: $\sqrt{7+2} - 1 \stackrel{?}{=} \sqrt{7-3}$

$\sqrt{9} - 1 \stackrel{?}{=} \sqrt{4}$

$3 - 1 = 2$

The solution set is {7}.

29. $\sqrt{3x+1} + \sqrt{2x+4} = 3$

$\sqrt{3x+1} = 3 - \sqrt{2x+4}$

$3x+1 = 9 - 6\sqrt{2x+4} + 2x + 4$ Square both sides.

$3x+1 = -6\sqrt{2x+4} + 2x + 13$

$x-12 = -6\sqrt{2x+4}$

$x^2-24x+144 = 36(2x+4)$ Square both sides again.

$x^2-24x+144 = 72x+144$

$x^2-96x = 0$

$x(x-96) = 0$

$x = 0$ or $x-96 = 0$

$x = 0$ or $x = 96$

Check: $\sqrt{3(0)+1} + \sqrt{2(0)+4} \stackrel{?}{=} 3$

$\sqrt{1} + \sqrt{4} \stackrel{?}{=} 3$

$1 + 2 = 3$

$\sqrt{3(96)+1} + \sqrt{2(96)+4} \stackrel{?}{=} 3$

$\sqrt{289} + \sqrt{196} \stackrel{?}{=} 3$

$17 + 14 \neq 3$

The solution set is {0}.

33. $\sqrt{1+2\sqrt{x}} = \sqrt{x+1}$

$1+2\sqrt{x} = x+1$ Square both sides.

$2\sqrt{x} = x$

$4x = x^2$ Square both sides again.

$0 = x^2-4x$

$0 = x(x-4)$

$x = 0$ or $x-4 = 0$

$x = 0$ or $x = 4$

Check: $\sqrt{1+2\sqrt{0}} \stackrel{?}{=} \sqrt{0+1}$

$\sqrt{1} = \sqrt{1}$

$\sqrt{1+2\sqrt{4}} \stackrel{?}{=} \sqrt{4+1}$

$\sqrt{1+4} = \sqrt{4+1}$

The solution set is {0,4}.

37. $2n^4-9n^2+4 = 0$

$(2n^2-1)(n^2-4) = 0$

$2n^2-1 = 0$ or $n^2-4 = 0$

$2n^2 = 1$ or $n^2 = 4$

$n^2 = \frac{1}{2}$ or $n = \pm 2$

$n = \pm\frac{\sqrt{2}}{2}$ or $n = \pm 2$

The solution set is $\{\pm \frac{\sqrt{2}}{2}, \pm 2\}$.

41. $x^4-4x^2+1 = 0$

Let $y = x^2$; then the equation becomes

$y^2-4y+1 = 0$.

Using the quadratic formula we obtain

$$y = \frac{4 \pm \sqrt{16-4}}{2} = \frac{4 \pm \sqrt{12}}{2} = \frac{4 \pm 2\sqrt{3}}{2} = 2 \pm \sqrt{3}.$$

Therefore, $x^2 = 2 + \sqrt{3}$ or $x^2 = 2 - \sqrt{3}$

$x = \pm\sqrt{2+\sqrt{3}}$ or $x = \pm\sqrt{2-\sqrt{3}}$

The solution set is

$\{\pm\sqrt{2+\sqrt{3}}, \pm\sqrt{2-\sqrt{3}}\}$.

45. $6x^{\frac{2}{3}}-5x^{\frac{1}{3}}-6 = 0$

$(2x^{\frac{1}{3}}-3)(3x^{\frac{1}{3}}+2) = 0$

$2x^{\frac{1}{3}}-3 = 0$ or $3x^{\frac{1}{3}}+2 = 0$

$2x^{\frac{1}{3}} = 3$ or $3x^{\frac{1}{3}} = -2$

$x^{\frac{1}{3}} = \frac{3}{2}$ or $x^{\frac{1}{3}} = -\frac{2}{3}$

$\left(x^{\frac{1}{3}}\right)^3 = (\frac{3}{2})^3$ or $\left(x^{\frac{1}{3}}\right)^3 = (-\frac{2}{3})^3$

$x = \frac{27}{8}$ or $x = -\frac{8}{27}$

The solution set is $\{-\frac{8}{27}, \frac{27}{8}\}$.

<u>Check</u>: $6\left(\frac{27}{8}\right)^{\frac{2}{3}} - 5\left(\frac{27}{8}\right)^{\frac{1}{3}} - 6 \stackrel{?}{=} 0$

$6(\frac{9}{4}) - 5(\frac{3}{2}) - 6 \stackrel{?}{=} 0$

$\frac{27}{2} - \frac{15}{2} - \frac{12}{2} \stackrel{?}{=} 0$

$0 = 0$

$6\left(-\frac{8}{27}\right)^{\frac{2}{3}} - 5\left(-\frac{8}{27}\right)^{\frac{2}{3}} - 6 \stackrel{?}{=} 0$

$6(\frac{4}{9}) - 5(-\frac{2}{3}) - 6 \stackrel{?}{=} 0$

$\frac{8}{3} + \frac{10}{3} - \frac{18}{3} \stackrel{?}{=} 0$

$0 = 0$

49. $x-11\sqrt{x}+30 = 0$

Let $y = \sqrt{x}$; then the equation becomes

$y^2-11y+30 = 0$.

$(y-6)(y-5) = 0$

$y-6 = 0$ or $y-5 = 0$

$y = 6$ or $y = 5$

Now substitute $\sqrt{x}$ for y and solve.

$\sqrt{x} = 6$ or $\sqrt{x} = 5$

$x = 36$ or $x = 25$

The solution set is $\{25,36\}$.

<u>Check</u>: $36-11\sqrt{36}+30 \stackrel{?}{=} 0$

$36-11(6)+30 \stackrel{?}{=} 0$

$36 - 66 + 30 \stackrel{?}{=} 0$

$0 = 0$

$25-11\sqrt{25}+30 \stackrel{?}{=} 0$

$25-11(5)+30 \stackrel{?}{=} 0$

$25 - 55 + 30 \stackrel{?}{=} 0$

$0 = 0$

Problem Set 2.6

9. $-2x+1 > 5$

$2x > 4$

$-\frac{1}{2}(-2x) < -\frac{1}{2}(4)$ Don't forget to reverse the inequality since we are multiplying both sides by a negative number.

$x < -2$

The solution set is $(-\infty,-2)$.

13. $6(2t-5)-2(4t-1) \geq 0$

$12t-30-8t+2 \geq 0$

$4t-28 \geq 0$

$4t \geq 28$

$\frac{1}{4}(4t) \geq \frac{1}{4}(28)$

$t \geq 7$

The solution set is $[7,\infty)$.

17. $\frac{n+2}{4} + \frac{n-3}{8} < 1$

$8(\frac{n+2}{4} + \frac{n-3}{8}) < 8(1)$

$2(n+2)+n-3 < 8$

$2n+4+n-3 < 8$

$3n+1 < 8$

$3n < 7$

$n < \frac{7}{3}$

The solution set is $(-\infty,\frac{7}{3})$.

21. $.09x+.1(x+200) > 77$

$100[.09x+.1(x+200)] > 100(77)$

$9x+10(x+200) > 7700$

$9x+10x+2000 > 7700$

$19x > 5700$

$x > 300$

The solution set is $(300,\infty)$.

25. $3 \geq \frac{7-x}{2} \geq 1$

$2(3) \geq 2(\frac{7-x}{2}) \geq 2(1)$ Multiply through by 2.

$6 \geq 7-x \geq 2$

$-1 \geq -x \geq -5$

$1 \leq x \leq 5$ Multiply through by -1.

The solution set is $[1,5]$.

29. $x^2-2x-15 > 0$

$(x-5)(x+3) > 0$

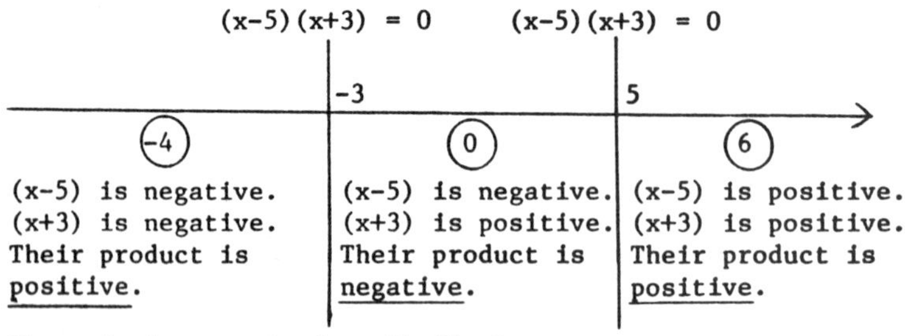

The solution set is $(-\infty,-3)\cup(5,\infty)$.

33. $3t^2+11t-4 > 0$

$(3t-1)(t+4) > 0$

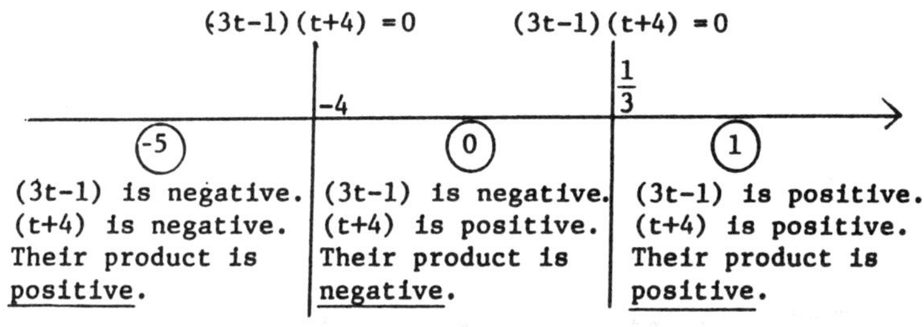

The solution set is $(-\infty,-4)\cup(\frac{1}{3},\infty)$.

37. $4x^2-4x+1 > 9$

$(2x-1)(2x-1) > 0$

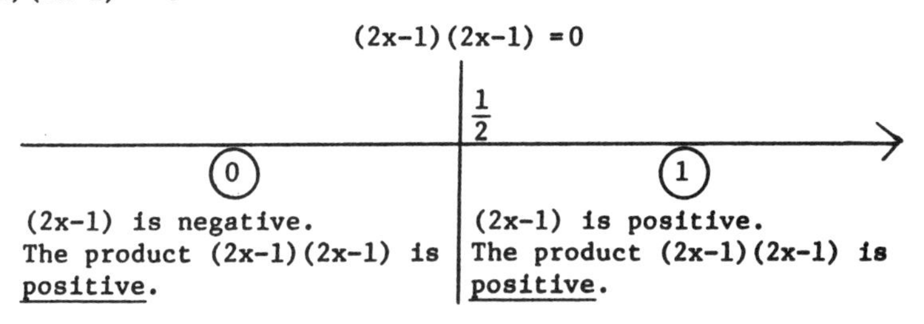

The solution set is $(-\infty,\frac{1}{2})\cup(\frac{1}{2},\infty)$.

You may also recognize the solution set from the statement $(2x-1)^2 > 0$. Squaring any number, except 0, produces a positive result.

41. $(x+1)(x-2) \geq (x-4)(x+6)$

$$x^2-x-2 \geq x^2+2x-24$$

$$-3x \geq -22$$

$$x \leq \frac{22}{3}$$

The solution set is $(-\infty,\frac{22}{3}]$.

45. $(x+2)(2x-1)(x-5) \le 0$

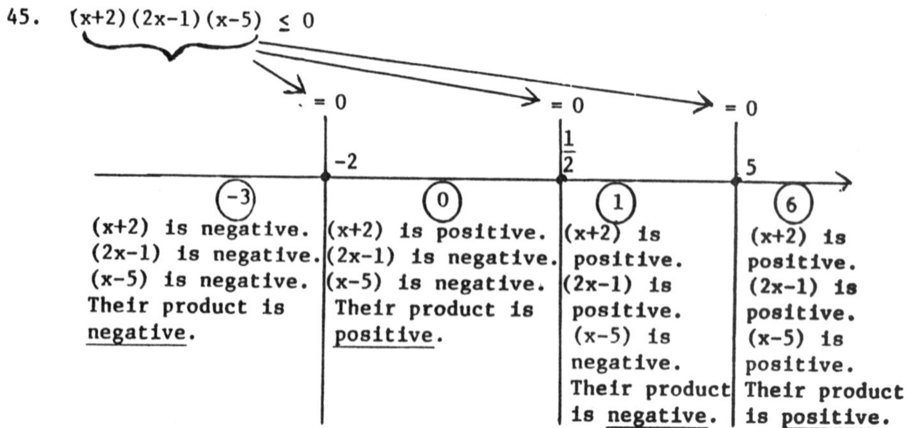

The solution set is $(-\infty,-2]\cup[\frac{1}{2},5]$.

49. $(x-2)^2(x+3) > 0$

$(x-2)^2(x+3) = 0$ $\quad$ $(x-2)^2(x+3) = 0$

-3 $\quad$ 2

(-4): $(x-2)^2$ is positive. $(x+3)$ is negative. The product $(x-2)^2(x+3)$ is negative.

(0): $(x-2)^2$ is positive. $(x+3)$ is positive. The product $(x-2)^2(x+3)$ is positive.

(3): $(x-2)^2$ is positive. $(x+3)$ is positive. The product $(x-2)^2(x+3)$ is positive.

The solution set is $(-3,2)\cup(2,\infty)$.

53. Let x be her score on the 5th exam.

$$\frac{94+84+86+88+x}{5} \ge 90$$

$$\frac{352+x}{5} \ge 90$$

$$352+x \ge 450$$

$$x \ge 98$$

She has to get a 98 or better on the 5th exam.

57.

$$80 \le \frac{100M}{11} \le 140$$

$$880 \le 100M \le 1540$$

$$8.8 \le M \le 15.4$$

Their mental ages are between 8.8 and 15.4, inclusive.

Problem Set 2.7

1. $\frac{x+1}{x-5} > 0$

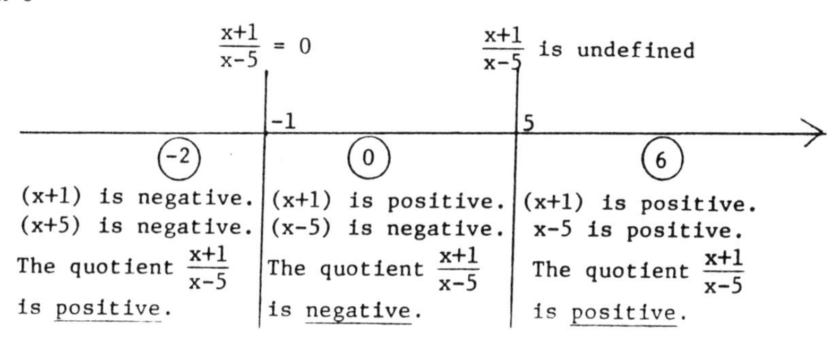

The solution set is $(-\infty,-1)\cup(5,\infty)$.

5. $\frac{-x+3}{3x-1} \geq 0$

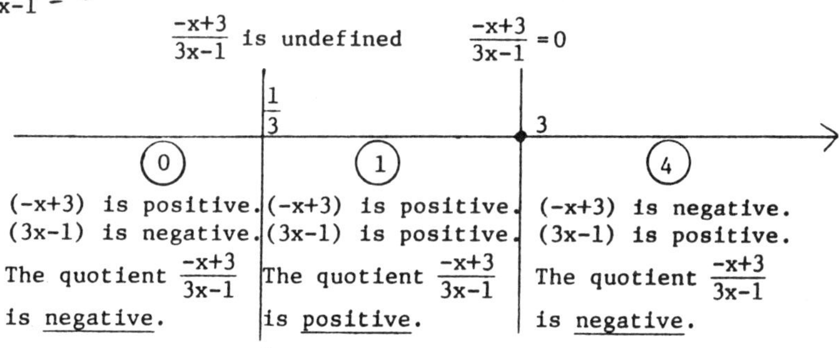

The solution set is $(\frac{1}{3},3]$.

9.

$$\frac{x-1}{x+2} < 2$$

$$\frac{x-1}{x+2} - 2 < 0$$

$$\frac{x-1-2(x+2)}{x+2} < 0$$

$$\frac{x-1-2x-4}{x+2} < 0$$

$$\frac{-x-5}{x+2} < 0$$

$\frac{-x-5}{x+2} = 0$ at -5; $\frac{-x-5}{x+2}$ is undefined at -2.

Test point -6: $(-x-5)$ is positive. $(x+2)$ is negative. The quotient $\frac{-x-5}{x+2}$ is negative.

Test point -4: $(-x-5)$ is negative. $(x+2)$ is negative. The quotient $\frac{-x-5}{x+2}$ is positive.

Test point 0: $(-x-5)$ is negative. $(x+2)$ is positive. The quotient $\frac{-x-5}{x+2}$ is negative.

The solution set is $(-\infty,-5)\cup(-2,\infty)$.

13. $\frac{1}{x-2} < \frac{1}{x+3}$

$\frac{1}{x-2} - \frac{1}{x+3} < 0$

$\frac{x+3-(x-2)}{(x-2)(x+3)} < 0$

$\frac{x+3-x+2}{(x-2)(x+3)} < 0$

$\frac{5}{(x-2)(x+3)} < 0$

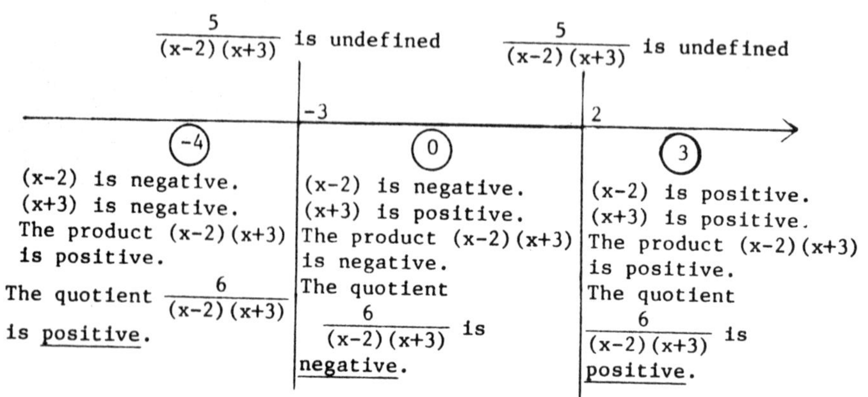

The solution set is $(-3,2)$.

17. $\left|x + \frac{1}{4}\right| = \frac{2}{5}$

$x + \frac{1}{4} = -\frac{2}{5}$ or $x + \frac{1}{4} = \frac{2}{5}$

$x = -\frac{2}{5} - \frac{1}{4}$ or $x = \frac{2}{5} - \frac{1}{4}$

$x = -\frac{13}{20}$ or $x = \frac{3}{20}$

The solution set is $\{-\frac{13}{20}, \frac{3}{20}\}$.

21. $|3x+4| = 5$

$3x+4 = -5$ or $3x+4 = 5$

$3x = -9$ or $3x = 1$

$x = -3$ or $x = \frac{1}{3}$

The solution set is $\{-3,\frac{1}{3}\}$.

25. $|-3x-2| = 8$

$-3x-2 = -8$ or $-3x-2 = 8$

$-3x = -6$ or $-3x = 10$

$x = 2$ or $x = -\frac{10}{3}$

The solution set is $\{-\frac{10}{3},2\}$.

29. $|3x-1| = |2x+3|$

$3x-1 = -(2x+3)$ or $3x-1 = 2x+3$

$3x-1 = -2x-3$ or $x = 4$

$5x = -2$

$x = -\frac{2}{5}$

The solution set is $\{-\frac{2}{5},4\}$.

33. $|x-2| = |x+4|$

$x-2 = -(x+4)$ or $x-2 = x+4$

$x-2 = -x-4$ or $0 = 6$

$2x = -2$

$x = -1$

The solution set is $\{-1\}$.

37. By applying Property 2.7, $|x| > 8$ is equivalent to $x < -8$ or $x > 8$. Therefore, the solution set is $(-\infty,-8)\cup(8,\infty)$.

41. $|t-3| > 5$

$t-3 < -5$ or $t-3 > 5$

$t < -2$ or $t > 8$

The solution set is $(-\infty,-2)\cup(8,\infty)$.

45. $|3n+2| > 9$

$3n+2 < -9$ or $3n+2 > 9$

$3n < -11$ or $3n > 7$

$n < -\frac{11}{3}$ or $n > \frac{7}{3}$

The solution set is $(-\infty,-\frac{11}{3})\cup(\frac{7}{3},\infty)$.

49. $|3-2x| < 4$

$-4 < 3-2x < 4$

$-7 < -2x < 1$

$\frac{7}{2} > x > -\frac{1}{2}$

The solution set is $(-\frac{1}{2},\frac{7}{2})$.

53. $|-2-x| \leq 5$

$-5 \leq -2-x \leq 5$

$-3 \leq -x \leq 7$

$3 \geq x \geq -7$

The solution set is $[-7,3]$.

57. $|x+4|-1 > 1$

$|x+4| > 2$

$x+4 < -2$ or $x+4 > 2$

$x < -6$ or $x > -2$

The solution set is $(-\infty,-6)\cup(-2,\infty)$.

61. $-2|x+1| > -10$

$|x+1| < 5$

$-5 < x+1 < 5$

$-6 < x < 4$

The solution set is $(-6,4)$.

65. $\left|\frac{x-1}{x+3}\right| > 1$

$\frac{x-1}{x+3} < -1$ or $\frac{x-1}{x+3} > 1$

$\frac{x-1}{x+3}+1 < 0$ or $\frac{x-1}{x+3}-1 > 0$

$\frac{x-1+x+3}{x+3} < 0$ or $\frac{x-1-(x+3)}{x+3} > 0$

$\frac{2x+2}{x+3} < 0$ or $\frac{-4}{x+3} > 0$

Using a number line and test numbers, each of these inequalities can be solved.

$\frac{2x+2}{x+3} < 0 \longrightarrow (-3,-1)$

$\frac{-4}{x+3} > 0 \longrightarrow (-\infty,-3)$

Therefore, the solution of $\frac{2x+2}{x+3} < 0$ or $\frac{-4}{x+3} > 0$ and consequently of the original inequality is $(-\infty,-3)\cup(-3,-1)$.

69. $\left|\frac{k}{2k-1}\right| \le 2$

$$\frac{k}{2k-1} \ge -2 \text{ and } \frac{k}{2k-1} \le 2$$

$$\frac{k}{2k-1} + 2 \ge 0 \quad \text{and} \quad \frac{k}{2k-1} - 2 \le 0$$

$$\frac{k+2(2k-1)}{2k-1} \ge 0 \text{ and } \frac{k-2(2k-1)}{2k-1} \le 0$$

$$\frac{k+4k-2}{2k-1} \ge 0 \text{ and } \frac{k-4k+2}{2k-1} \le 0$$

$$\frac{5k-2}{2k-1} \ge 0 \text{ and } \frac{-3k+2}{2k-1} \le 0$$

Using a number line and test numbers, we can solve the two inequalities.

$$\frac{5k-2}{2k-1} \ge 0 \longrightarrow (-\infty,\tfrac{2}{5}]\cup(\tfrac{1}{2},\infty)$$

$$\frac{-3k+2}{2k-1} \le 0 \longrightarrow (-\infty,\tfrac{1}{2})\cup[\tfrac{2}{3},\infty)$$

Therefore, the solution set of $\frac{5k-2}{2k-1} \ge 0$ and $\frac{-3k+2}{2k-1} \le 0$ and consequently of the original inequality is $(-\infty,\frac{2}{5}]\cup[\frac{2}{3},\infty)$.

CHAPTER 3

Problem Set 3.1

1. $6-(-4) = 10$

5. $|4-(-2)| = |6| = 6$

9. $1 + \frac{2}{3}(10-1) = 1 + \frac{2}{3}(9) = 1+6 = 7$

13. $-1 + \frac{3}{5}(-11-(-1)) = -1 + \frac{3}{5}(-10) = -1+(-6) = -7$

17. $AB = \sqrt{(3-1)^2+(-4-(-1))^2} = \sqrt{2^2+(-3)^2} = \sqrt{13}$

midpoint of AB is $(\frac{1+3}{2}, \frac{-1+(-4)}{2}) = (2,-\frac{5}{2})$

21. $AB = \sqrt{(-\frac{1}{3}-\frac{1}{2})^2+(\frac{3}{2}-\frac{1}{3})^2} = \sqrt{(-\frac{5}{6})^2+(\frac{7}{6})^2} = \sqrt{\frac{74}{36}} = \frac{\sqrt{74}}{6}$

midpoint of AB is $\left(\frac{\frac{1}{2}+(-\frac{1}{3})}{2}, \frac{\frac{1}{3}+\frac{3}{2}}{2}\right) = (\frac{1}{12},\frac{11}{12})$

25. $x = -2 +\frac{2}{5}(8-(-2)) = -2 +\frac{2}{5}(10) = -2+4 = 2$

$y = 1 +\frac{2}{5}(11-1) = 1 +\frac{2}{5}(10) = 1+4 = 5$

29. (a) The midpoint of the segment determined by (2,4) and (10,13) is

$(\frac{2+10}{2}, \frac{4+13}{2}) = (6,\frac{17}{2})$.

Then the midpoint of the segment determined by (2,4) and $(6,\frac{17}{2})$ is

$\left(\frac{2+6}{2}, \frac{4+\frac{17}{2}}{2}\right) = (4,\frac{25}{4})$.

(b) $x = 2 +\frac{1}{4}(10-2) = 2 +\frac{1}{4}(8) = 2+2 = 4$

$y = 4 +\frac{1}{4}(13-4) = 4 +\frac{1}{4}(9) = 4 +\frac{9}{4} = \frac{25}{4}$

The point is $(4,\frac{25}{4})$.

33.

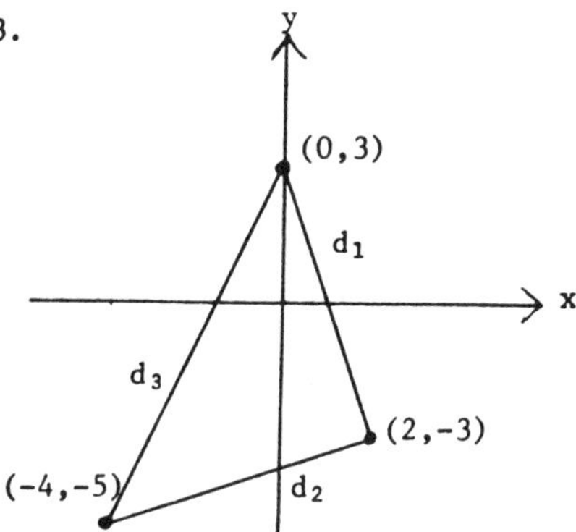

$d_1 = \sqrt{(0-2)^2+(3-(-3))^2} = \sqrt{40}$

$d_2 = \sqrt{(2-(-4))^2+(-3-(-5))^2} = \sqrt{40}$

$d_3 = \sqrt{(0-(-4))^2+(3-(-5))^2} = \sqrt{80}$

$d_1 = d_2$

37. Name the points A(4,-5),B(6,7),C(-8,-3), and D(-1,2).

$$AD = \sqrt{(4-(-1))^2+(-5-2)^2} = \sqrt{25+49} = \sqrt{74}$$

$$BD = \sqrt{(6-(-1))^2+(7-2)^2} = \sqrt{49+25} = \sqrt{74}$$

$$CD = \sqrt{(-8-(-1))^2+(-3-2)^2} = \sqrt{49+25} = \sqrt{74}$$

41. The ratio $\frac{AP}{PB} = \frac{2}{1}$ means that P is two-thirds of the way from A to B. Therefore,

$$x = -1 + \frac{2}{3}(5-(-1)) = -1 + \frac{2}{3}(6) = -1+4 = 3$$

$$y = 2 + \frac{2}{3}(11-2) = 2 + \frac{2}{3}(9) = 2+6 = 8.$$

The point is (3,8).

Problem Set 3.2

For Problems 1-16, each graph is a straight line. Therefore, we can find two points and draw the line. Usually the two points involving the intercepts are easy to find and then it's a good idea to find a third point to serve as a check.

17. First, graph $x+2y = 4$ as a dashed line since equality is not included in the given statement. Then use a test point such as (0,0). The statement $0+2(0) > 4$ is false; therefore, the half-plane not containing the origin should be shaded.

21. First, graph $2x+5y = 10$ as a solid line because equality is included in the given statement. Then use (0,0) as a test point. The statement $2(0)+5(0) \leq 10$ is true; therefore, the half-plane containing the origin should be shaded.

25. First, graph $y = -x$ as a solid line because equality is included in the given statement. Since the origin is on the line, we must use some other point as a test point. Let's use (1,2). The statement $2 \leq -1$ is false; therefore, the half-plane not containing (1,2) should be shaded.

29. First, graph $x = -1$ as a dashed line because equality is not included in the given statement. Use (0,0) as a test point. The statement $0 > -1$ is true; therefore, the half-plane containing the origin should be shaded.

33. The equation $|x-y| = 2$ is equivalent to $x-y = 2$ or $x-y = -2$. Thus, the graph of $|x-y| = 2$ is the two lines $x-y = 2$ and $x-y = -2$.

37. The inequality $|2x+3y| > 6$ is equivalent to $2x+3y < -6$ or $2x+3y > 6$. Thus, the graph of $|2x+3y| > 6$ is as shown in the back of the text.

41. If $y \geq 0$, then $|y| = y$ and the given equation becomes $y = x$. If $y < 0$, then $|y| = -y$ and the given equation becomes $-y = x$ which is equivalent to $y = -x$.

Problém Set 3.3

1. Let (3,1) be P_1 and (7,4) be P_2.

$$m = \frac{y_2-y_1}{x_2-x_1} = \frac{4-1}{7-3} = \frac{3}{4}$$

5. Let (-4,2) be P_1 and (-2,2) be P_2.

$$m = \frac{y_2-y_1}{x_2-x_1} = \frac{2-2}{-2-(-4)} = \frac{0}{2} = 0$$

9. $$\frac{6-4}{x-(-2)} = \frac{2}{9}$$

$$\frac{2}{x+2} = \frac{2}{9}$$

$$\begin{aligned} 2(x+2) &= 18 \\ 2x+4 &= 18 \\ 2x &= 14 \\ x &= 7 \end{aligned}$$

13. A slope of $\frac{2}{3}$ means that for every 2 units of vertical change there must be 3 units of horizontal change between points on the line. For example,

$m = \frac{2}{3}$, (3,2) $\xrightarrow[\text{units to the right produces}]{\text{2 units up and 3}}$ (6,4)

$m = \frac{4}{6}$, (3,2) $\xrightarrow[\text{units to the right produces}]{\text{4 units up and 6}}$ (9,6)

$m = \frac{-2}{-3}$, (3,2) $\xrightarrow[\text{units to the left produces}]{\text{2 units down and 3}}$ (0,0)

17. $m = -\frac{3}{5} = \frac{-3}{5}$, (2,-1) $\xrightarrow[\text{units to the right produces}]{\text{3 units down and 5}}$ (7,-4)

$m = -\frac{3}{5} = \frac{3}{-5}$, (2,-1) $\xrightarrow[\text{units to the left produces}]{\text{3 units up and 5}}$ (-3,2)

$m = -\frac{3}{5} = \frac{-6}{10}$, (2,-1) $\xrightarrow[\text{units to the right produces}]{\text{6 units down and 10}}$ (12,-7)

21. Remember that a problem of this type can be done by using the general approach described in the text or by substituting into the point-slope form. We shall illustrate both techniques for this problem.

General Approach

Choose a point (x,y) to represent any other point on the line. The slope determined by (x,y) and (-1,-2) is to be 2.

$$\frac{y+2}{x+1} = 2$$

$$\begin{aligned} 2x+2 &= y+2 \\ 2x-y &= 0 \end{aligned}$$

Point-Slope Form

Substitute 2 for m, -1 for x, and -2 for y_1 in the point-slope form.

$$y-y_1 = m(x-x_1)$$

$$\begin{aligned} y-(-2) &= 2(x-(-1)) \\ y+2 &= 2(x+1) \\ y+2 &= 2x+2 \\ 0 &= 2x-y \end{aligned}$$

25. Using the point-slope form, we obtain

$$y-y_1 = m(x-x_1)$$

$$\begin{aligned} y-(-2) &= 0(x-5) \\ y+2 &= 0 \\ y &= -2 \end{aligned}$$

29. The points (-1,7) and (5,2) determine a slope of $\frac{2-7}{5-(-1)} = \frac{-5}{6} = -\frac{5}{6}$. Now using this slope and one of the two given points, we can substitute into the point-slope form.

$$y-y_1 = m(x-x_1)$$
$$y-7 = -\frac{5}{6}(x-(-1)) \qquad \text{(We used the point (-1,7).)}$$
$$6y-42 = -5x-5$$
$$5x+6y = 37$$

33. The points (4,-3) and (-7,-3) determine a slope of $\frac{-3-(-3)}{-7-4} = \frac{0}{-11} = 0$. Then using one of the points and the slope we can substitute into the point-slope form.

$$y-y_1 = m(x-x_1)$$
$$y+3 = 0(x-4)$$
$$y = -3$$

37. This problem can be done by using (a) a general approach, (b) the point-slope form (b = 2 means that the point (0,2) is on the line), or (c) the slope-intercept form. Using the slope-intercept form, we obtain

$$y = mx+b$$
$$y = -\frac{3}{7}x+2.$$

41. Substituting into the slope-intercept form, we obtain

$$y = mx+b$$
$$y = -\frac{5}{6}x+\frac{1}{4}.$$

45. Lines parallel to the y-axis have equations of the form $x = a$. Thus, the equation of the line is $x = -4$.

49. By changing to slope-intercept form, we can find the slope of the given line.

$$x-4y = 7$$
$$-4y = -x+7$$
$$y = \frac{1}{4}x-\frac{7}{4}$$

The slope of the given line is $\frac{1}{4}$. Therefore, the slope of the line perpendicular to it is -4. Now using the point (-2,6) and $m = -4$, we can substitute into the point-slope form.

$$y-y_1 = m(x-x_1)$$
$$y-6 = -4(x+2)$$
$$y-6 = -4x-8$$
$$4x+y = -2$$

53. Let's change each equation to slope-intercept form.

$$5x-7y = 14 \qquad 7x+5y = 12$$
$$-7y = -5x+14 \qquad 5y = -7x+12$$
$$y = \frac{5}{7}x - 2 \qquad y = -\frac{7}{5}x + \frac{12}{5}$$

Their slopes, $\frac{5}{7}$ and $-\frac{7}{5}$, are negative reciprocals of each other. Therefore, the lines are perpendicular.

57. Change each equation to slope-intercept form.

$$x+y = 0 \qquad x-y = 0$$
$$y = -x \qquad -y = -x$$
$$y = x$$

The two slopes, -1 and 1, are negative reciprocals of each other; therefore, the lines are perpendicular.

61. $x-2y = 7$

$-2y = -x+7$

$y = \frac{1}{2}x - \frac{7}{2}$

$m = \frac{1}{2} \qquad b = -\frac{7}{2}$

65. $7x-5y = 12$

$-5y = -7x+12$

$y = \frac{7}{5}x - \frac{12}{5}$

$m = \frac{7}{5} \qquad b = -\frac{12}{5}$

69. Label the points A(6,6), B(2,-2), C(-8,-5), and D(-4,3).

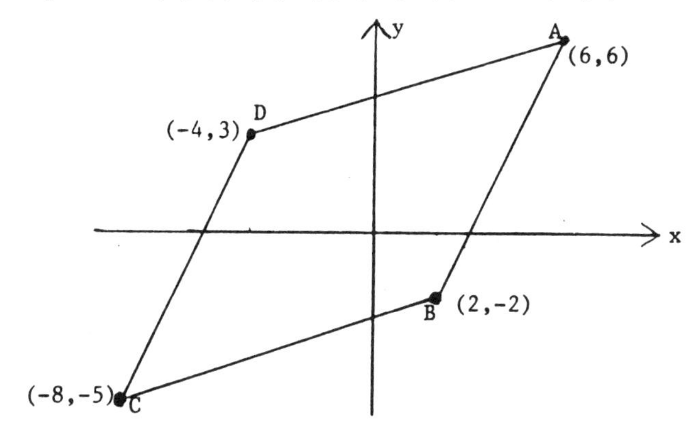

slope of $\overline{DA}$ is $\frac{6-3}{6-(-4)} = \frac{3}{10}$

slope of $\overline{CB}$ is $\frac{-5-(-2)}{-8-2} = \frac{-3}{-10} = \frac{3}{10}$

Therefore, $\overline{DA}$ is parallel to $\overline{CB}$.

slope of $\overline{DC}$ is $\frac{-5-3}{-8-(-4)} = \frac{-8}{-4} = 2$

slope of $\overline{BA}$ is $\frac{-2-6}{2-6} = \frac{-8}{-4} = 2$

Therefore, $\overline{DC}$ is parallel to $\overline{BA}$ and the figure is a parallelogram.

73.

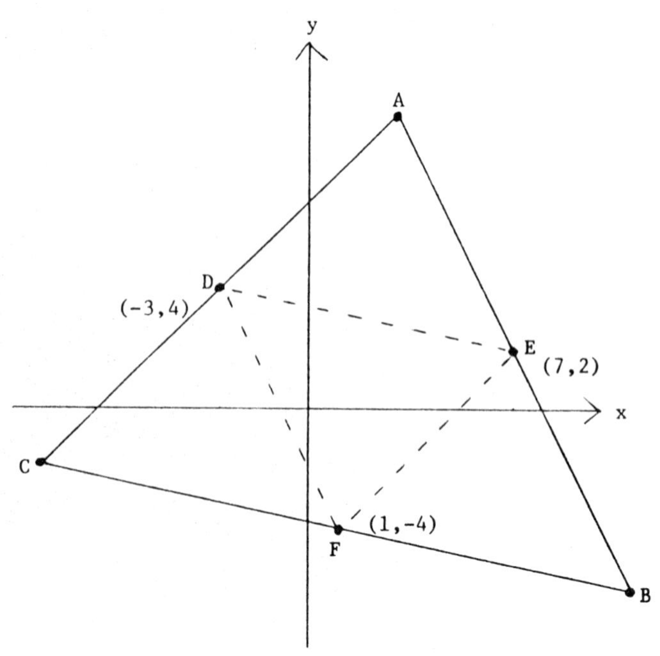

$\overline{CB}$ must be parallel to $\overline{DE}$; so let's find the slope of $\overline{DE}$.

$$\text{slope of } \overline{DE} = \frac{2-4}{7-(-3)} = \frac{-2}{10} = -\frac{1}{5}$$

Therefore, the equation of $\overline{CB}$ is

$$y+4 = -\frac{1}{5}(x-1)$$

$$5y+20 = -x+1$$
$$x+5y = -19.$$

$\overline{AC}$ must be parallel to $\overline{EF}$; so let's find the slope of $\overline{EF}$.

$$\text{slope of } \overline{EF} = \frac{2-(-4)}{7-1} = \frac{6}{6} = 1$$

Therefore, the equation of $\overline{AC}$ is

$$y-4 = 1(x+3)$$
$$y-4 = x+3$$
$$-7 = x-y.$$

$\overline{AB}$ must be parallel to $\overline{DF}$; so let's find the slope of $\overline{DF}$.

$$\text{slope of } \overline{DF} = \frac{4-(-4)}{-3-1} = \frac{8}{-4} = -2$$

Therefore, the equation of $\overline{AB}$ is

$$y-2 = -2(x-7)$$
$$y-2 = -2x+14$$
$$2x+y = 16.$$

77. (a) Let x represent the measure of the run.

$$\frac{3}{5} = \frac{19}{x}$$
$$3x = 95$$
$$x = 31\frac{2}{3}$$

To the nearest centimeter, the run is 32 centimeters.

Problem Set 3.4

1.

This point is symmetric to given point with respect to the y-axis.

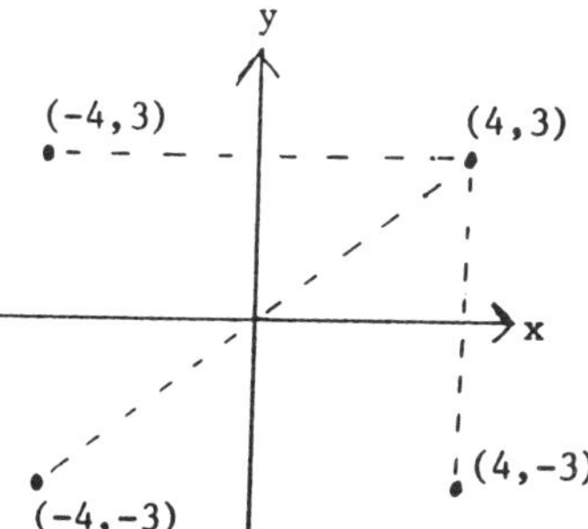

This point is symmetric to given point with respect to the origin.

This point is symmetric to given point with respect to the x-axis.

5. 1. Replace y with -y to obtain the point symmetric to the given point with respect to the x-axis. Therefore, the x-axis reflection of (0,4) is (0,-4).

2. Replace x with -x to obtain the point symmetric to the given point with respect to the y-axis. Therefore, the y-axis reflection of (0,4) is (0,4).

3. Replace x with -x and y with -y to obtain the point symmetric to the given point with respect to the origin. Therefore, the origin reflection of (0,4) is (0,-4).

9. Replacing y with -y produces an equivalent equation.

$$x^3 = y^2$$
$$x^3 = (-y)^2$$
$$x^3 = y^2$$

Therefore, the equation exhibits x-axis symmetry.

13 and 17. None of the tests for symmetry are satisfied; therefore, there is no symmetry with respect to either axis nor the origin.

21. Replacing x with -x produces an equivalent equation. Therefore, the graph has y-axis symmetry.

If x = 0, then y = 0; so the origin is on the graph.

Because of y-axis symmetry, we can limit our table of values to positive values for x.

x	y
1	1
2	4
3	9

Plotting these points along with (0,0), connecting them with a smooth curve, and reflecting this portion of the curve across the y-axis produces the figure indicated in the answer section of the text.

25. Replacing x with -x and y with -y produces an equivalent equation. Therefore, the graph has origin symmetry.

Neither x nor y can equal zero; so the graph does not have points on either axis.

Solving the given equation for y produces $y = \frac{4}{x}$ and this form is convenient for obtaining a table of values such as the following.

x	y
$\frac{1}{2}$	8
1	4
2	2
3	$\frac{4}{3}$
4	1

Plotting these points, connecting them with a smooth curve, and refelcting this portion of the curve through the origin produces the figure indicated in the answer section of the text.

29. Replacing y with -y produces an equation equivalent to the given equation. Therefore, the graph has x-axis symmetry. If $x = 0$, then $y = 0$; so the origin is a part of the graph. Solving the given equation for y produces $y = \pm\sqrt{x^3}$. Using $y = \sqrt{x^3}$, a table of values can be constructed.

x	y
1	1
2	$2\sqrt{2}$
3	$3\sqrt{3}$

Plotting these points along with (0,0), connecting them with a smooth curve, and reflecting this part of the graph across the x-axis produces the figure indicated in the answer section of this text.

33. None of the tests for symmetry are satisfied.

If $x = 0$, then $y = 0$; so the origin is a part of the graph.

Because of $\sqrt{x}$, x must be nonnegative.

Using some positive values for x, (0 has already been used) the following table of values can be formed.

x	y
1	-1
4	-2
9	-3

Plotting these points, along with (0,0), and connecting them with a smooth curve produces the figure indicated in the answer section of the text.

37. Replacing x with -x produces an equivalent equation and replacing y with -y produces an equivalent equation. Therefore, the graph is symmetric with respect to both axes and the origin.

If $x = 0$, then $0+2y^2 = 8$

$$y^2 = 4$$

$$y = \pm 2.$$

So the points (0,2) and (0,-2) are on the graph.

If $y = 0$, then $x^2 = 8$

$$x = \pm\sqrt{8} = \pm 2\sqrt{2}.$$

So the points $(2\sqrt{2},0)$ and $(-2\sqrt{2},0)$ are on the graph.

Solving for y produces

$$x^2+y^2 = 8$$

$$2y^2 = 8-x^2$$

$$y^2 = \frac{8-x^2}{2}$$

$$y = \pm\sqrt{\frac{8-x^2}{2}}$$

Using $y = \sqrt{\frac{8-x^2}{2}}$ and some positive values for x produces the following table.

x	y
1	$\frac{\sqrt{14}}{2}$
2	$\sqrt{2}$
$\sqrt{2}$	$\sqrt{3}$

Plotting these points along with (0,2) and $(2\sqrt{2},0)$, connecting them with a smooth curve, and reflecting this part of the graph across the x-axis, across the y-axis, and through the origin produces the figure indicated in the answer section of the text.

41. None of the tests for symmetry are satisfied. If x = 2, then y = 0; so the point (2,0) is on the graph. The points (3,1),(6,2),(11,3) along with (2,0) should determine the graph.

45. Replacing y with -y produces $x = (-y)^2+2 = y^2+2$; so the graph has x-axis symmetry. If y = 0, then x = 2; so the point (2,0) is on the graph. Additional points (3,1) and (6,2) can be determined. By plotting (2,0), (3,1) and (6,2) and then reflecting across the x-axis we should obtain the figure shown in the answer section of the text.

Problem Set 3.5

For Problems 1-23, the general form of the equation of a circle, $(x-h)^2+(y-k)^2 = r^2$, can be used.

1. Substitute 2 for h, 3 for k, and 5 for r into the general form and simplify.

$$(x-2)^2+(y-3)^2 = 5$$
$$x^2-4x+4+y^2-6y+9 = 25$$
$$x^2+y^2-4x-6y-12 = 0$$

5. Substitute 3 for h, 0 for k, and 3 for r into the general form and simplify.

$$(x-3)^2+(y-0)^2 = 3$$
$$x^2-6x+9+y^2 = 9$$
$$x^2+y^2-6x = 0$$

For Problems 9, 13, and 17 we will change the equation to the general form by completing the square on x and y.

9. $$x^2+y^2-6x-10y+30 = 0$$
$$x^2-6x+\_\_ +y^2-10y+\_\_ = -30$$
$$x^2-6x+9+y^2-10y+25 = -30+9+25$$
$$(x-3)^2+(y-5)^2 = 2^2 \quad (4 = 2^2)$$

Therefore, h = 3, k = 5, and r = 2

13. $$x^2+y^2-10x = 0$$
$$x^2-10x+\_\_ +y^2 = 0$$
$$x^2-10x+25+y^2 = 0+25$$
$$(x-5)^2+(y-0)^2 = 5^2 \quad (25 = 5^2)$$

Therefore, h = 5, k = 0, and r = 5.

17. $$4x^2+4y^2-4x-8y-11 = 0$$

$$x^2+y^2-x-2y-\frac{11}{4} = 0$$ Divide both sides by 4.

$$x^2-x+ \_\_ +y^2-2y+ \_\_ = \frac{11}{4}$$

$$x^2-x+\frac{1}{4}+y^2-2y+1 = \frac{11}{4}+\frac{1}{4}+1$$

$$\left(x-\frac{1}{2}\right)^2+(y-1)^2 = 2^2 \quad \left(\frac{11}{4}+\frac{1}{4}+1 = 4 = 2^2\right)$$

Therefore, $h = \frac{1}{2}$, $k = 1$, and $r = 2$.

21. Sketching a figure helps with the analysis of this problem.

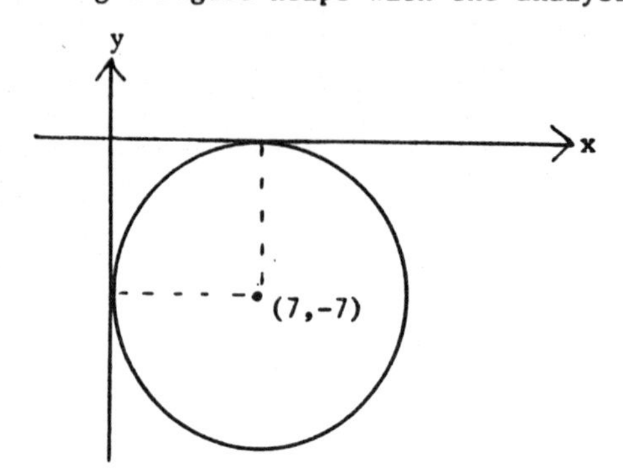

The center is at (7,-7) and $r = 7$. Therefore, substitute 7 for h, -7 for k, and 7 for r and simplify.

$$(x-7)^2+(y+7)^2 = 7^2$$

$$x^2-14x+49+y^2+14y+49 = 49$$

$$x^2+y^2-14x+14y+49 = 0$$

25. If $x = 0$, then $4y^2 = 36$

$$y^2 = 9$$

$$y = \pm 3.$$

The points (0,3) and (0,-3) are the endpoints of the major axis of the ellipse.

If $y = 0$, then $9x^2 = 36$

$$x^2 = 4$$

$$x = \pm 2.$$

The points (2,0) and (-2,0) are the endpoints of the minor axis of the ellipse.

29. The form of the equation indicates that it is a circle. So let's complete the square.

$$x^2+y^2-4x = 0$$
$$x^2-4x+\_\_ +y^2 = 0$$
$$x^2-4x+4+y^2 = 0+4$$
$$(x-2)^2+(y-0)^2 = 2^2$$

The graph is a circle with its center at (2,0) and having a radius of 2 units.

33. If $x = 0$, then $y = \pm 3$.

The points (0,3) and (0,-3) are vertices of the hyperbola.

We can find the equations of the asymptotes by replacing the constant term with 0 and solving for y.

$$y^2-3x^2 = 0$$
$$y^2 = 3x^2$$
$$y = \pm\sqrt{3}\,x.$$

Thus, the equations of the asymptotes are $y = \sqrt{3}\,x$ and $y = -\sqrt{3}\,x$.

37. If $x = 0$, then $3y^2 = 12$

$$y^2 = 4$$
$$y = \pm 2.$$

Thus, the points (0,2) and (0,-2) are the endpoints of the major axis of the ellipse.

If $y = 0$, then $4x^2 = 12$

$$x^2 = 3$$
$$x = \pm\sqrt{3}.$$

Thus, the points $(\sqrt{3},0)$ and $(-\sqrt{3},0)$ are the endpoints of the minor axis of the ellipse.

41. Replacing x with -x and y with -y produces an equivalent equation. Thus, the graph has origin symmetry.

Since $x \neq 0$ and $y \neq 0$, no points of either axis are on the graph.

Solving $xy = 2$ for y produces $y = \frac{2}{x}$ and this equation, along with positive values for x, helps determine the following table.

x	y
$\frac{1}{2}$	4
1	2
2	1
3	$\frac{2}{3}$

Plotting these points, connecting them with a smooth curve, and reflecting this portion of the curve through the origin produces the hyperbola indicated in the answer section of the text.

45. If $xy = 0$, then $x = 0$ or $y = 0$. Thus, the graph consists of the x-axis and the y-axis.

CHAPTER 4

Problem Set 4.1

1. Using $f(x) = -2x+5$, we can determine the following.

$$f(3) = -2(3)+5 = -6+5 = -1$$
$$f(5) = -2(5)+5 = -10+5 = -5$$
$$f(-2) = -2(-2)+5 = 4+5 = 9$$

5. Using $h(x) = \frac{2}{3}x - \frac{3}{4}$, we can determine the following.

$$h(3) = \frac{2}{3}(3) - \frac{3}{4} = 2 - \frac{3}{4} = \frac{8}{4} - \frac{3}{4} = \frac{5}{4}$$

$$h(4) = \frac{2}{3}(4) - \frac{3}{4} = \frac{8}{3} - \frac{3}{4} = \frac{32}{12} - \frac{9}{12} = \frac{23}{12}$$

$$h(-\frac{1}{2}) = \frac{2}{3}(-\frac{1}{2}) - \frac{3}{4} = -\frac{1}{3} - \frac{3}{4} = -\frac{4}{12} - \frac{9}{12} = -\frac{13}{12}$$

9. Use $f(x) = x$ to find $f(4)$ and $f(10)$.

$f(4) = 4$ and $f(10) = 10$

Use $f(x) = x^2$ to find $f(-3)$ and $f(-5)$.

$f(-3) = (-3)^2 = 9$ and $f(-5) = (-5)^2 = 25$

13. Use $f(x) = 1$ to find $f(2)$.

$f(2) = 1$

Use $f(x) = 0$ to find $f(0)$ and $f(-\frac{1}{2})$.

$f(0) = 0$ and $f(-\frac{1}{2}) = 0$

Use $f(x) = -1$ to find $f(-4)$.

$f(-4) = -1$

17. $f(a+h) = -(a+h)^2+4(a+h)-2 = -a^2-2ha-h^2+4a+4h-2$

$f(a) = -a^2+4a-2$

Therefore,

$$\frac{f(a+h)-f(a)}{h} = \frac{-a^2-2ha-h^2+4a+4h-2+a^2-4a+2}{h} = \frac{-2ha-h^2+4h}{h}$$
$$= \frac{h(-2a-h+4)}{h} = -2a-h+4$$

21. $f(a+h) = (a+h)^3-(a+h)^2+2(a+h)-1 = a^3+3a^2h+3ah^2+h^3-a^2-2ha-h^2+2a+2h-1$

$f(a) = a^3-a^2+2a-1$

Therefore,

$$\frac{f(a+h)-f(a)}{h} = \frac{a^3+3a^2h+3ah^2+h^3-a^2-2ha-h^2+2a+2h-1-a^3+a^2-2a+1}{h}$$
$$= \frac{3a^2h+3ah^2+h^3-2ha-h^2+2h}{h}$$
$$= \frac{h(3a^2+3ah+h^2-2a-h+2)}{h} = 3a^2+3ah+h^2-2a-h+2$$

25. $f(a+h) = \dfrac{1}{(a+h)^2}$ and $f(a) = \dfrac{1}{a^2}$

Therefore,

$$\frac{f(a+h)-f(a)}{h} = \frac{\dfrac{1}{(a+h)^2} - \dfrac{1}{a^2}}{h}$$

$$= \frac{\dfrac{a^2-(a+h)^2}{a^2(a+h)^2}}{h} = \frac{a^2-(a^2+2ha+h^2)}{ha^2(a+h)^2} = \frac{a^2-a^2-2ha-h^2}{ha^2(a+h)^2}$$

$$= \frac{-h(2a+h)}{ha^2(a+h)^2} = -\frac{2a+h}{a^2(a+h)^2}$$

29. It is not a function because some vertical lines would intersect the graph in more than one point.

33. It is a function because no vertical line would intersect the graph in more than one point.

37. The domain of $f(x) = x^2-2$ is the set of all real numbers since any real number can be substituted for x. The smallest value that f(x) takes on is at $x = 0$. Thus, the range is all real numbers greater than or equal to -2.

41. The domain of $f(x) = -\sqrt{x}$ is the set of all nonnegative real numbers since $-\sqrt{x}$ is a real number for only those values. Since $f(x) = -\sqrt{x}$ is non-positive for all values of x, the range is the set of all nonpositive real numbers.

45. Since the denominator of the given function cannot equal zero, we must discard $x = \frac{1}{2}$ and $x = -4$. Thus, the domain is $\{x|x \neq \frac{1}{2}$ and $x \neq -4\}$.

49. We need to eliminate values of x that make x^2-x-12 equal to zero.

$$x^2-x-12 = 0$$
$$(x-4)(x+3) = 0$$
$$x-4 = 0 \text{ or } x+3 = 0$$
$$x = 4 \text{ or } x = -3$$

Thus, the domain is $\{x|x \neq 4$ and $x \neq -3\}$.

53. The radicand, x^2-16, must be nonnegative.

$$x^2-16 \geq 0$$
$$(x+4)(x-4) \geq 0$$

Using the number line approach from Section 2.6, we find that $(x+4)(x-4) \geq 0$ is satisfied by $x \geq 4$ or $x \leq -4$. Thus, using interval notation the domain is the set $(-\infty,-4]\cup[4,\infty)$.

57. The radicand, $x^2-3x-40$, must be nonnegative.

$$x^2-3x-40 \geq 0$$
$$(x-8)(x+5) \geq 0$$

Using the number line approach from Section 2.6, we find that $(x+8)(x+5) \geq 0$ is satisfied by $x \geq 8$ or $x \leq -5$. Thus, using interval notation the domain is the set $(-\infty,-5]\cup[8,\infty)$.

61. $A(2) = \Pi(2)^2 = 4\Pi = 12.57$

$A(3) = \Pi(3)^2 = 9\Pi = 28.27$ (nearest hundredth)

$A(12) = \Pi(12)^2 = 144\Pi = 452.39$

$A(17) = \Pi(17)^2 = 289\Pi = 907.92$

65. Using $I(r) = 500r$ we obtain

$I(.11) = \$500(.11) = \$55,$

$I(.12) = \$500(.12) = \$60,$

$I(.135) = \$500(.135) = \$67.50,$

$I(.15) = \$500(.15) = \$75.$

69. Since $f(x) = x^3$ and $f(-x) = (-x)^3 = -x^3$, we see that $f(-x) = -f(x)$. Therefore, $f(x) = x^3$ is an odd function.

73. Since $f(x) = x^3+1$ and $f(-x) = (-x)^3+1 = -x^3+1$, we see that $f(-x) \neq f(x)$ and $f(-x) \neq -f(x)$. Therefore, the function $f(x) = x^3+1$ is neither even nor odd.

77. Since $f(x) = x^5+x^3+x$ and $f(-x) = -x^5-x^3-x$, we see that $f(-x) = -f(x)$. Therefore, the function $f(x) = x^5+x^3+x$ is odd.

Problem Set 4.2

1. Solve: $2x-4 = 0$
   $2x = 4$
   $x = 2$

   So, $f(2) = 0$ and the point $(2,0)$ is on the line.

   Furthermore, $f(0) = 2(0)-4 = -4$; so the point $(0,-4)$ is on the line.

   The two points $(2,0)$ and $(0,-4)$ determine the line.

5. $f(0) = -2(0) = 0$; so the point $(0,0)$ is on the line.

   $f(1) = -2(1) = -2$; so the point $(1,-2)$ is on the line.

   The two points $(0,0)$ and $(1,-2)$ determine the line.

9. $f(0) = -1$ and $f(2) = -1$. The two points $(0,-1)$ and $(2,-1)$ determine the horizontal line.

13. Remember that any equation of the form $f(x) = ax^2$ is a parabola with its vertex at the origin. It opens upward if $\underline{a}$ is positive and downward if $\underline{a}$ is negative. Using $f(1) = 3$ and $f(-1) = 3$, along with the origin, allows us to sketch the parabola as indicated in the answer section of the text.

17. $f(x) = (x+2)^2 = (x-(-2))^2$. So this is the basic parabola shifted 2 units to the left.

21. This is the basic parabola moved 1 unit to the right and 2 units up. So the vertex is at $(1,2)$ and two other points, $(0,3)$ and $(2,3)$, can be used to help determine the width.

25. $f(x) = x^2+2x+4$
$= x^2+2x+1+4-1$
$= (x+1)^2+3$

This is the basic parabola moved 1 unit to the left and 3 units up.

29. $f(x) = 2x^2+12x+17$
$= 2(x^2+6x+ \_\_)+17$
$= 2(x^2+6x+9)+17-18$
$= 2(x+3)^2-1$

This is the parabola $f(x) = 2x^2$ moved 3 units to the left and 1 unit down.

33. $f(x) = 2x^2-2x+3$
$= 2(x^2-x+ \_\_)+3$
$= 2(x^2-x+\frac{1}{4})+3-\frac{1}{2}$
$= 2(x-\frac{1}{2})^2+\frac{5}{2}$

This is the parabola $f(x) = 2x^2$ moved $\frac{1}{2}$ unit to the right and $\frac{5}{2}$ units up.

37. For nonnegative values of x, the function to be graphed is $f(x) = x$.
For negative values of x, the function to be graphed is $f(x) = 3x$.

41. For nonnegative values of x, the function to be graphed is $f(x) = 2$.
For negative values of x, the function to be graphed is $f(x) = -1$.

45. Note that for all values of x such that $0 \leq x < 1$, $f(x) = 0$. For example, $f(0) = 0$, $f(\frac{1}{4}) = 0$, $f(\frac{1}{2}) = 0$, and $f(\frac{3}{4}) = 0$. Therefore, for $0 \leq x < 1$, the graph of $f(x) = [x]$ is the horizontal segment $f(x) = 0$. Likewise, for $1 \leq x < 2$, $f(x) = 1$. Therefore, for $1 \leq x < 2$, the graph of $f(x) = [x]$ is the horizontal segment $f(x) = 1$. The completed graph is shown in the answer section of the text.

Problem Set 4.3

1. Step 1: Because $a = 1$, the parabola opens upward.

Step 2: $-\frac{b}{2a} = -\frac{-8}{2(1)} = 4$

Step 3: $f(-\frac{b}{2a}) = f(4) = 16-32+15 = -1$

Thus, the vertex is at $(4,-1)$.

Step 4: Letting $x = 5$, we obtain $f(5) = 0$. Thus, $(5,0)$ is on the graph and so is its reflection $(3,0)$ across the line of symmetry $x = 4$.

The three points $(4,-1)$, $(5,0)$, and $(3,0)$ are used to sketch the parabola.

5. Step 1: Because $a = -1$, the parabola opens downward.

Step 2: $-\frac{b}{2a} = -\frac{4}{2(-1)} = 2$

Step 3: $f(-\frac{b}{2a}) = f(2) = -4+8-7 = -3$

Thus, the vertex is at $(2,-3)$.

Step 4: Letting $x = 3$, we obtain $f(3) = -9+12-7 = 4$. So $(3,-4)$ is on the graph and so is its reflection $(1,-4)$ across the line of symmetry $x = 2$.

The three points $(2,-3)$, $(3,-4)$, and $(1,-4)$ are used to sketch the parabola.

9. Step 1: Because a = 1, the parabola opens upward.

Step 2: $-\frac{b}{2a} = -\frac{3}{2(1)} = -\frac{3}{2}$

Step 3: $f(-\frac{b}{2a}) = f(-\frac{3}{2}) = \frac{9}{4} - \frac{9}{2} - 1 = -\frac{13}{4}$

Thus, the vertex is at $(-\frac{3}{2}, -\frac{13}{4})$.

Step 4: Letting x = 0, we obtain f(0) = -1. So (0,-1) is on the graph and so is its reflection (-3,-1) across the line of symmetry $x = -\frac{3}{2}$.

The three points $(-\frac{3}{2}, -\frac{13}{4})$, (0,-1), and (-3,-1) are used to sketch the parabola.

13. The graph of $f(x) = -x^2+3$ is the graph of $f(x) = -x^2$ moved up 3 units.

17. $f(x) = -2x^2+4x+1$

$= -2(x^2-2x+\_\_\ )+1$

$= -2(x^2-2x+1)+1+2$

$= -2(x-1)^2+3$

This is the graph of $f(x) = -2x^2$ moved 1 unit to the right and 3 units up.

21. To find the x-intercepts, let f(x) = 0 and solve the resulting equation.

$x^2-8x+15 = 0$
$(x-5)(x-3) = 0$
$x-5 = 0$ or $x-3 = 0$
$x = 5$ or $x = 3$

To find the vertex, let's complete the square.

$f(x) = x^2-8x+15$
$= x^2-8x+16+15-16$
$= (x-4)^2-1$

Therefore, the vertex is at (4,-1).

25. To find the x-intercepts, let f(x) = 0 and solve the resulting equation.

$-x^2+10x-24 = 0$
$x^2-10x+24 = 0$
$(x-4)(x-6) = 0$
$x-4 = 0$ or $x-6 = 0$
$x = 4$ or $x = 6$

To find the vertex, let's determine the point $(-\frac{b}{2a}, f(-\frac{b}{2a}))$.

$f(x) = -x^2+10x-24$

$-\frac{b}{2a} = -\frac{10}{2(-1)} = -\frac{10}{-2} = 5$

$f(5) = -5^2+10(5)-24 = 1$

Therefore, the vertex is at (5,1).

29. To find the x-intercepts, let f(x) = 0 and solve the resulting equation.

$-x^2+9x-21 = 0$
$x^2-9x+21 = 0$

$$x = \frac{9\pm\sqrt{81-84}}{2}$$

At this stage we recognize that the solutions for x are nonreal complex numbers. Thus, this parabola does not intersect the x-axis.

To find the vertex, let's determine the point $(-\frac{b}{2a}, f(-\frac{b}{2a}))$.

$$-\frac{b}{2a} = -\frac{9}{2(-1)} = \frac{9}{2}$$

$$f(\frac{9}{2}) = -(\frac{9}{2})^2 + 9(\frac{9}{2}) - 21$$

$$= -\frac{81}{4} + \frac{81}{2} - \frac{84}{4}$$

$$= \frac{-81+162-84}{4} = -\frac{3}{4}$$

Therefore, the vertex is at $(\frac{9}{2}, -\frac{3}{4})$.

33. The function will be maximized at the vertex of the parabola.

$$-\frac{b}{2a} = -\frac{280}{2(-2)} = 70$$

They need to sell 70 items.

37. Let x represent one number and 50-x the other number. Then the product function is

$$f(x) = x(50-x) = -x^2+50x.$$

This function is minimized at the vertex of the parabola.

$$-\frac{b}{2a} = \frac{50}{2(-1)} = 25$$

The two numbers are x = 25 and 50-x = 50-25 = 25.

41. Let x be the number of \$.25 decreases in the monthly rate. Therefore, 1000+20x represents the new number of subscribers and \$15 - .25x represents what each subscriber must pay per month. Therefore, the revenue function is

$$f(x) = (1000+20x)(15-.25x) = -5x^2+50x+15000.$$

The function is maximized at the vertex of the parabola.

$$-\frac{b}{2a} = -\frac{50}{2(-5)} = 5$$

So 5 decreases of \$.25 each amounts to \$1.25. The new rate will be \$15 - 1.25 = \$13.75 and there should be 1000 + 20x = 1000 +20(5) = 1100 subscribers.

45.

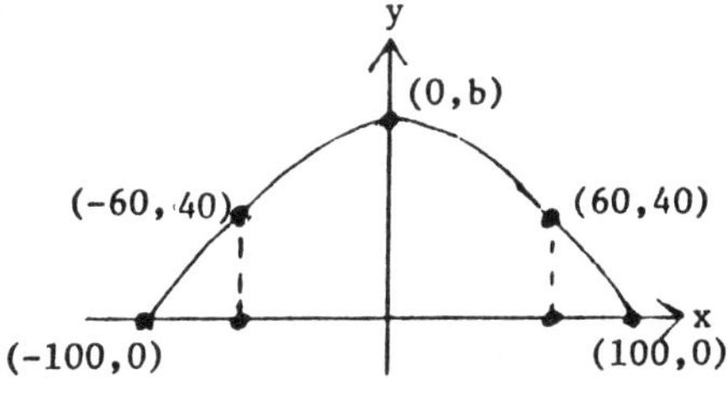

The equation of the parabola is of the form $f(x) = ax^2+b$. Using the points (60,40) and (100,0), we can determine the following.

$f(60) = 3600a+b = 40$
$f(100) = 10000a+b = 0$

Solving the system $\begin{pmatrix} 3600a+b = 40 \\ 10000a+b = 0 \end{pmatrix}$ produces $b = 62.5$.

The arch must be 62.5 feet high above the stream.

Problem Set 4.4

1. This is the graph of $f(x) = x^4$ shifted upward two units.

5. This is the x-axis reflection of $f(x) = x^3$.

9. This is the basic absolute value graph shifted one unit to the right and two units upward.

13. If $x \geq 0$, then $|x| = x$ and the function becomes $f(x) = x+|x| = x+x = 2x$. So for nonnegative values of x, we graph the function $f(x) = 2x$.

    If $x < 0$, then $|x| = -x$ and the given function becomes $f(x) = x+|x| = x-x = 0$. So for negative values of x, we graph the function $f(x) = 0$ that is, the x-axis.

17. If $x \geq 0$, then $|x| = x$ and the function becomes $f(x) = x-|x| = x-x = 0$. So for nonnegative values of x, we graph $f(x) = 0$.

    If $x < 0$, then $|x| = -x$ and the function becomes $f(x) = x-|x| = x-(-x) = 2x$. So for negative values of x, we graph $f(x) = 2x$.

21. This is the graph of $f(x) = \sqrt{x}$ shifted two units to the left and three units downward.

25. This is the graph of $f(x) = x^4$ stretched by a factor of two reflected through the x-axis, and shifted upward one unit.

29. This is the graph of $f(x) = x^3$ stretched by a factor of three, shifted two units to the right, and shifted downward one unit.

Problem Set 4.5

1. $(f+g)x = f(x)+g(x) = (3x-4)+(5x+2) = 8x-2$

   $(f-g)x = f(x)-g(x) = (3x-4)-(5x+2) = -2x-6$

   $(f\cdot g)x = f(x)\cdot g(x) = (3x-4)(5x+2) = 15x^2-14x-8$

   $(\frac{f}{g})x = \frac{f(x)}{g(x)} = \frac{3x-4}{5x+2}$

5. $(f+g)x = f(x)+g(x) = (x^2-x-1)+(x^2+4x-5) = 2x^2+3x-6$

   $(f-g)x = f(x)-g(x) = (x^2-x-1)-(x^2+4x-5) = -5x+4$

   $(f\cdot g)x = f(x)\cdot g(x) = (x^2-x-1)(x^2+4x-5) = x^4+3x^3-10x^2+x+5$

   $(\frac{f}{g})x = \frac{f(x)}{g(x)} = \frac{x^2-x-1}{x^2+4x-5}$

9. $(f\circ g)x = f(g(x)) = f(3x-1) = 2(3x-1) = 6x-2$

   $(g\circ f)x = g(f(x)) = g(2x) = 3(2x)-1 = 6x-1$

13. $(f\circ g)x = f(g(x)) = f(x^2+1) = 3(x^2+1)+4 = 3x^2+7$

    $(g\circ f)x = g(f(x)) = g(3x+4) = (3x+4)^2+1 = 9x^2+24x+16+1 = 9x^2+24x+17$

17. $(f \circ g)x = f(g(x)) = f(2x+7) = \frac{1}{2x+7}$

$(g \circ f)x = g(f(x)) = g(\frac{1}{x}) = 2(\frac{1}{x})+7 = \frac{2}{x} + 7 = \frac{2+7x}{x}$

21. $(f \circ g)x = f(g(x)) = f(\frac{2}{x}) = \frac{1}{\frac{2}{x} - 1} = \frac{1}{\frac{2-x}{x}} = \frac{x}{2-x}$

$(g \circ f)x = g(f(x)) = g(\frac{1}{x-1}) = \frac{2}{\frac{1}{x-1}} = 2(x-1) = 2x-2$

25. $(f \circ g)(x) = f(\frac{x+1}{x}) = \frac{1}{\frac{x+1}{x} - 1} = \frac{1}{\frac{x+1-x}{x}} = x$

$(g \circ f)(x) = g(\frac{1}{x-1}) = \frac{\frac{1}{x-1} + 1}{\frac{1}{x-1}} = \frac{\frac{1+x-1}{x-1}}{\frac{1}{x-1}} = x$

29. $(f \circ g)(-2) = f(g(-2)) = f((-2)^2-3(-2)-4) = f(6) = 2(6)-3 = 9$

$(g \circ f)(1) = g(f(1)) = g(2(1)-3) = g(-1) = (-1)^2-3(-1)-4 = 0$

33. $f(g(x)) = f(\frac{1}{2}x) = 2(\frac{1}{2}x) = x$

$g(f(x)) = g(2x) = \frac{1}{2}(2x) = x$

37. $f(g(x)) = f(\frac{x-4}{3}) = 3(\frac{x-4}{3})+4 = x-4+4 = x$

$g(f(x)) = g(3x+4) = \frac{3x+4-4}{3} = \frac{3x}{3} = x$

41. $f(g(x)) = f(ax+b) = 3(ax+b)-4 = 3ax+3b-4$

$g(f(x)) = g(3x-4) = a(3x-4)+b = 3ax-4a+b$

Now we can equate $f(g(x))$ and $g(f(x))$.

$$3ax+3b-4 = 3ax-4a+b$$
$$2b+4a = 4$$
$$b+2a = 2$$

Problem Set 4.6

1 and 5. Yes, it is a one-to-one function because no horizontal line intersects the graph in more than one point.

9. Since $f(x_1) \neq f(x_2)$ when $x_1 \neq x_2$, we know that the function $f(x) = x^3$ is one-to-one.

13. Note that $f(2) = -16$ and $f(-2) = -16$. Therefore, for different values of x, the functional values are equal, so it is not a one-to-one function.

17. Remember that the domain of a function is the set of all first components of the ordered pairs. The range of a function is the set of all second components of the ordered pairs. An inverse function can be formed by exchanging the components of each ordered pair.

21. $f(g(x)) = f(-2x + \frac{5}{3}) = -\frac{1}{2}(-2x + \frac{5}{3}) + \frac{5}{6} = x - \frac{5}{6} + \frac{5}{6} = x$

$g(f(x)) = g(-\frac{1}{2}x + \frac{5}{6}) = -2(-\frac{1}{2}x + \frac{5}{6}) + \frac{5}{3} = x - \frac{5}{3} + \frac{5}{3} = x$

25. $f(g(x)) = f(\frac{x^2+4}{2}) = \sqrt{2(\frac{x^2+4}{2})-4} = \sqrt{x^2+4-4} = \sqrt{x^2} = x$

$g(f(x)) = g(\sqrt{2x-4} = (\frac{\sqrt{2x-4}^2+4}{2}) = \frac{2x-4+4}{2} = \frac{2x}{2} = x$

29. $f(g(x)) = f(\sqrt[3]{x}) - (\sqrt[3]{x})^3 = x$

$g(f(x)) = g(x^3) = \sqrt[3]{x^3} = x$

Yes, they are inverse functions.

33. $f(g(x)) = f(\sqrt{x+3}) = (\sqrt{x+3})^2-3 = x+3-3 = x$

$g(f(x)) = g(x^2-3) = \sqrt{x^2-3+3} + \sqrt{x^2} = x$ Yes, they are inverse functions.

37. Let $y = x-4$; then exchange variables to form $x = y-4$ and solve for y.

$y = x+4$

Therefore, $f^{-1}(x) = x+4$.

41. Let $y = \frac{3}{4}x - \frac{5}{6}$; then exchange variables to obtain $x = \frac{3}{4}y - \frac{5}{6}$ and solve for y.

$$x = \frac{3}{4}y - \frac{5}{6}$$

$$12x = 9y-10$$

$$12x+16 = 9y$$

$$\frac{12x+10}{9} = y$$

45. First, restrict x such that $x \geq 0$. Then let $y = \sqrt{x}$, exchange variables to form $x = \sqrt{y}$, and solve for y.

$$x = \sqrt{y}$$

$$x^2 = y$$

Therefore, $f^{-1}(x) = x^2$ for $x \geq 0$.

49. Let $y = 1 + \frac{1}{x}$ for $x > 0$. Interchange variables and solve for y.

$$x = 1 + \frac{1}{y} \text{ for } y > 0$$

$$x-1 = \frac{1}{y}$$

$$\frac{1}{x-1} = y \text{ for } x > 1$$

(Note that $x > 1$ will insure $y > 0$.)

Thus, the inverse is $f^{-1}(x) = \frac{1}{x-1}$ for $x > 1$.

53. Let $y = 2x+1$ and interchange variables to obtain $x = 2y+1$. Now we can solve for y in terms of x.

$$x = 2y+1$$
$$x-1 = 2y$$
$$\frac{x-1}{2} = y$$

Therefore, $f^{-1}(x) = \frac{x-1}{2}$.

57. Let $y = x^2-4$ and interchange the variables to produce $x = y^2-4$. Now solve for y in terms of x.

$$x = y^2-4$$
$$x+4 = y^2$$
$$\sqrt{x+4} = y \quad \text{for } x \geq -4$$

Therefore, $f^{-1}(x) = \sqrt{x+4}$ for $x \geq -4$.

61. By looking at the graph of $f(x) = -3x+1$, we see that the function is decreasing on $(-\infty,\infty)$.

65. By looking at the graph of $f(x) = -2x^2-16x-35$, we see that the function is increasing on $(-\infty,-4]$ and decreasing on $[-4,\infty)$.

Problem Set 4.7

9. Substitute 72 for y and 3 for x in the equation $y = kx$.

$$72 = k(3)$$
$$24 = k$$

13. Substitute 81 for A, 9 for b, and 18 for h in the equation $A = kbh$.

$$81 = k(9)(18)$$
$$81 = 162k$$
$$\frac{81}{162} = k$$
$$\frac{1}{2} = k$$

17. Substitute 18 for y, 9 for x, and 3 for w in the equation $y = \frac{kx^2}{w^3}$.

$$18 = \frac{k(9)^2}{(3)^3}$$
$$18 = \frac{81k}{27}$$
$$18 = 3k$$
$$6 = k$$

21. Substitute 96 for V, 36 for B, and 8 for h in the general equation $V = kBh$ to determine k.

$$96 = k(36)(8)$$
$$96 = 288k$$
$$\frac{96}{288} = k$$
$$\frac{1}{3} = k$$

Thus, the specific equation is $V = \frac{1}{3}Bh$. Now substitute 48 for B and 6 for h to determine V.

$$V = \frac{1}{3}(48)(6)$$

$$V = 96$$

25. Substitute 12 for ℓ and 4 for t in the general equation $t = k\sqrt{\ell}$ to determine k.

$$4 = k(\sqrt{12})$$

$$\frac{4}{\sqrt{12}} = k$$

$$\frac{2\sqrt{3}}{3} = k$$

Thus, the specific equation is $t = \frac{2\sqrt{3}}{3}\sqrt{\ell}$. Now substitute 3 for ℓ to determine t.

$$t = \frac{2\sqrt{3}}{3}(\sqrt{3}) = \frac{2(3)}{3} = 2 \text{ seconds}$$

29. Substitute 48 for V, 320 for T, and 20 for P in the general equation $V = \frac{kT}{P}$ to determine k.

$$48 = \frac{k(320)}{20}$$

$$48 = 16k$$

$$3 = k$$

Therefore, the specific equation is $V = \frac{3T}{P}$. Now substitute 280 for T and 30 for P to determine V.

$$V = \frac{3(280)}{30}$$

$$V = 28$$

33. (a) Substitute 385 for i, .11 for r, and 2 for t in the general equation $i = krt$ to determine k.

$$385 = k(.11)(2)$$

$$1750 = k$$

Therefore, the specific equation is $i = 1750rt$. Now substitute .12 for r and 1 for t to determine i.

$$i = 1750(.12)(1) = \$210$$

37. Substitute .08 for y and 225 for x in the general equation $y = \frac{k}{\sqrt{x}}$ to determine k.

$$.08 = \frac{k}{\sqrt{225}}$$

$$.08 = \frac{k}{15}$$

$$k = 1.2$$

Therefore, the specific equation is $y = \frac{1.2}{\sqrt{x}}$. Now substitute 625 for x to determine y.

$$y = \frac{1.2}{\sqrt{625}} = \frac{1.2}{25} = .048$$

CHAPTER 5

Problem Set 5.1

1. $3^x = 27$

$3^x = 3^3$

$x = 3$ If $b^n = b^m$, then $n = m$.

The solution set is $\{3\}$.

5. $3^{-x} = \frac{1}{81}$

$3^{-x} = \frac{1}{3^4}$

$3^{-x} = 3^{-4}$

$-x = -4$ If $b^n = b^m$, then $n = m$.

$x = 4$

The solution set is $\{4\}$.

9. $(\frac{2}{3})^t = \frac{9}{4}$

$(\frac{2}{3})^t = \frac{1}{\frac{4}{9}}$

$(\frac{2}{3})^t = \frac{1}{(\frac{2}{3})^2}$

$(\frac{2}{3})^t = (\frac{2}{3})^{-2}$

$t = -2$ If $b^n = b^m$, then $n = m$.

The solution set is $\{-2\}$.

13. $4^n = 8$

$(2^2)^n = 2^3$

$2^{2n} = 2^3$

$2n = 3$ If $b^n = b^m$, then $n = m$.

$n = \frac{3}{2}$

The solution set is $\{\frac{3}{2}\}$.

17. $(2^{2x-1})(2^{x+2}) = 32$

$2^{2x-1+x+2} = 2^5$

$2^{3x+1} = 2^5$

$3x+1 = 5$ If $b^n = b^m$, then $n = m$.

$3x = 4$

$x = \frac{4}{3}$

The solution set is $\{\frac{4}{3}\}$.

21. We know the general shape of exponential curves of the form $f(x) = b^x$ for $b > 1$. This knowledge, along with a few points such as $(0,1)$, $(1,3)$, $(2,9)$, $(-1,\frac{1}{3})$, and $(-2,\frac{1}{9})$, should determine this specific exponential curve.

25. We know the general shape of exponential curves of the form $f(x) = b^x$ where $0 < b < 1$. This knowledge, along with a few points such as $(0,1)$, $(1,\frac{2}{3})$, $(2,\frac{4}{9})$, $(-1,\frac{3}{2})$, and $(-2,\frac{9}{4})$ should determine this specific exponential curve.

29. This is the graph of $f(x) = 2^x$ moved 1 unit to the right.

33. Perhaps you will need to plot a few more points such as $(0,2)$, $(1,1)$, $(2,\frac{1}{2})$, $(3,\frac{1}{4})$, $(-1,4)$, and $(-2,8)$ to determine this graph.

37. $f(-x) = 3^{1-(x)^2} = 3^{1-x^2}$. Since $f(-x) = f(x)$, the graph is symmetric with respect to the vertical axis. Therefore, we can concentrate our point plotting using nonnegative values for x and then reflect across the vertical axis.

Problem Set 5.2

1. (a) Substitute \$.55 for P_0 and 3 for t in the equation $P = P_0(1.07)^t$.

$$P = .55(1.07)^3 = \$.67$$

5. $A = P(1 + \frac{r}{n})^{nt}$

$= 300(1 + \frac{.08}{2})^{2(6)}$

$= 300(1.04)^{12} = \$480.31$

9. $A = P(1 + \frac{r}{n})^{nt}$

$= 1000(1 + \frac{.12}{12})^{12(5)}$

$= 1000(1.01)^{60} = \$1816.70$

13. $A = P(1 + \frac{r}{n})^{nt} = 8000(1 + \frac{.105}{4})^{4(10)}$

$= 8000(1.02625)^{40} = \$22,553.65$

17. $A = Pe^{rt} = 750e^{.08(8)}$

$= 750e^{.64} = \$1422.36$

21. $A = Pe^{rt} = 7500e^{.085(10)}$

$= 7500e^{.85} = \$17,547.35$

25. $A = P(1 + \frac{r}{n})^{nt}$

$2700 = 1500(1 + \frac{r}{4})^{4(10)}$

$1.8 = (1 + \frac{r}{4})^{40}$ Divide both sides by 1500.

$1.014803164 \approx 1 + \frac{r}{4}$ Raise both sides to the $\frac{1}{40} = .025$ power.

$.014803164 \approx \frac{r}{4}$

$.0592126561 \approx r$

$r = 5.9\%$ to the nearest tenth of a percent

29. Let's determine the effective yield for each investment. The effective yield for the 8.25% compounded quarterly investment is calculated as follows.

$P(1+r) = P(1 + \frac{.0825}{4})^4$

$1+r = (1.020625)^4$

$r = (1.020625)^4-1$

$r \approx .0850876194$

$r = 8.51\%$ to the nearest hundredth of a percent

The effective yield of the 8.3% compounded semiannually investment can be calculated as follows.

$$P(1+r) = P(1 + \frac{.083}{2})^2$$

$$1+r = (1.0415)^2$$

$$r = (1.0415)^2 - 1$$

$$r \approx .08472225$$

$r = 8.47\%$ to the nearest hundredth of a percent

Therefore, the 8.25% compounded quarterly is the best investment.

33. $Q(t) = 1000e^{.4t}$

$$Q(2) = 1000e^{.4(2)} = 1000e^{.8} = 2226$$

$$Q(3) = 1000e^{.4(3)} = 1000e^{1.2} = 3320$$

$$Q(5) = 1000e^{.4(5)} = 1000e^{2} = 7389$$

(All answers have been rounded to the nearest whole number.)

37. (a) $P(a) = 14.7e^{-.21a}$

$$P(3.85) = 14.7e^{-.21(3.85)}$$

$$= 14.7e^{-.8085}$$

$= 6.5$ to the nearest tenth

41. The graph of $f(x) = 2e^x$ is the graph of $f(x) = e^x$ stretched by a factor of two. The points (0,2),(1,5.4), and (-1,.7) can be used to help sketch the curve.

Problem Set 5.3

17. Let $x = \log_5(\frac{1}{5})$; then $5^x = \frac{1}{5}$

$$5^x = 5^{-1}$$

$$x = -1$$

Therefore, $\log_5(\frac{1}{5}) = -1$.

21. Let $x = \log_3\sqrt{3}$; then $3^x = \sqrt{3}$

$$3^x = 3^{\frac{1}{2}}$$

$$x = \frac{1}{2}$$

Therefore, $\log_3\sqrt{3} = \frac{1}{2}$.

25. Let $x = \log_{\frac{1}{4}}(\frac{\sqrt[4]{32}}{2})$; then $(\frac{1}{4})^x = \frac{\sqrt[4]{32}}{2}$

$$(\frac{1}{2^2})^x = \frac{(32)^{\frac{1}{4}}}{2}$$

$$(2^{-2})^x = \frac{(2^5)^{\frac{1}{4}}}{2}$$

$$2^{-2x} = \frac{2^{\frac{5}{4}}}{2}$$

$$2^{-2x} = 2^{\frac{1}{4}}$$

$$-2x = \frac{1}{4}$$

$$x = -\frac{1}{8}$$

Therefore, $\log_{\frac{1}{4}}(\frac{\sqrt[4]{32}}{2}) = -\frac{1}{8}$.

29. $\log_5 5 = 1$ because $5^1 = 5$. Thus, $\log_2(\log_5 5) = \log_2 1$. Then let $x = \log_2 1$ and change to exponential form.

$$2^x = 1$$
$$2^x = 2^0$$
$$x = 0$$

Therefore, $\log_2(\log_5 5) = 0$.

33. Change $\log_8 t = \frac{5}{3}$ to exponential form and solve.

$$8^{\frac{5}{3}} = t$$
$$(\sqrt[3]{8})^5 = t$$
$$2^5 = t$$
$$32 = t$$

The solution set is $\{32\}$.

37. Change $\log_{10} x = 0$ to exponential form and solve.

$$10^0 = x$$
$$1 = x$$

The solution set is $\{1\}$.

41. $\log_2 125 = \log_2 5^3$
$= 3 \log_2 5$
$= 3(2.3219)$
$= 6.9657$

45. $\log_2 175 = \log_2(25 \cdot 7)$
$= \log_2 25 + \log_2 7$
$= \log_2 5^2 + \log_2 7$
$= 2 \log_2 5 + \log_2 7$
$= 2(2.3219) + 2.8074$
$= 7.4512$

49. $\log_8(\frac{5}{11}) = \log_8 5 - \log_8 11$
$= .7740 - 1.1531$
$= -.3791$

53. $\log_8 88 = \log (8 \cdot 11)$
$= \log_8 8 + \log_8 11$
$= 1 + 1.1531$
$= 2.1531$

57. Remember that $\log_b(xy) = \log_b x + \log_b y$ and this can be extended to the product of 3 or more numbers.

61. $\log_b\sqrt{xy} = \log_b(xy)^{\frac{1}{2}} = \log_b x^{\frac{1}{2}}y^{\frac{1}{2}}$

$= \log_b x^{\frac{1}{2}} + \log_b y^{\frac{1}{2}}$

$= \frac{1}{2}\log_b x + \frac{1}{2}\log_b y$

65. $\log_b x + \log_b y - \log_b z = \log_b(xy) - \log_b z$

$= \log_b(\frac{xy}{z})$

69. $\log_b x + \frac{1}{2}\log_b y = \log_b x + \log_b y^{\frac{1}{2}}$

$= \log_b x + \log_b\sqrt{y}$

$= \log_b(x\sqrt{y})$

73. $\log_3 x + \log_3 4 = 2$

$\log_3 4x = 2$

$4x = 3^2$ Change to exponential form.

$4x = 9$

$x = \frac{9}{4}$

The solution set is $\{\frac{9}{4}\}$.

77. $\log_2 x + \log_2(x-3) = 2$

$\log_2 x(x-3) = 2$

$x(x-3) = 2^2$ Change to exponential form.

$x(x-3) = 4$

$x^2-3x-4 = 0$

$(x-4)(x+1) = 0$

$x-4 = 0$ or $x+1 = 0$

$x = 4$ or $x = -1$

The solution set is {4}. The negative one solution is discarded since logarithms are only defined for positive numbers.

81. $\log_5(3x-2) = 1 + \log_5(x-4)$

$\log_5(3x-2)-\log_5(x-4) = 1$

$\log_5(\frac{3x-2}{x-4}) = 1$

$\frac{3x-2}{x-4} = 5^1$ Change to exponential form.

$3x-2 = 5x-20$

$18 = 2x$

$9 = x$

The solution set is {9}.

Problem Set 5.4

41\. (a) The points (.1,-1),(.5,-.3),(1,0),(2,.3),(4,.6),(8,.9), and (10,1) should determine this curve.

45\. The points $(-2,\frac{1}{9})$,$(-1,\frac{1}{3})$,(0,1),(1,3), and (2,9) should determine the graph of $g(x) = 3^x$. Then you can reflect each of these points through the line y = x and the resulting points will determine the graph of $f(x) = \log_3 x$.

49\. This is the graph of $f(x) = \log_2 x$ (Figure 5.6 in the text) shifted three units to the left.

53\. This is a stretching of the graph of $f(x) = \log_2 x$ by a factor of two. The points $(\frac{1}{2},-2)$,(1,0),(2,2),(4,4), and (8,6) will help determine the curve.

Problem Set 5.5

1\.

$$2^x = 9$$
$$\ln 2^x = \ln 9$$
$$x \ln 2 = \ln 9$$
$$x = \frac{\ln 9}{\ln 2} = 3.17 \quad \text{(nearest hundredth)}$$

(We can use either the natural logarithms or common logarithms since our calculators have both.

The solution set is {3.17}.

5\.

$$2^{x+1} = 7$$
$$\ln 2^{x+1} = \ln 7$$
$$(x+1)\ln 2 = \ln 7$$
$$x \ln 2 + \ln 2 = \ln 7$$
$$x \ln 2 = \ln 7 - \ln 2$$
$$x = \frac{\ln 7 - \ln 2}{\ln 2} = 1.81$$

The solution set is {1.81}.

9\.

$$e^x = 4.1$$
$$\ln e^x = \ln 4.1$$
$$x \ln e = \ln 4.1$$
$$x = \ln 4.1$$
$$x = 1.41$$

(The presence of e indicates that it would be better to use natural logarithms.)

(Remember that $\ln e = 1$.)

The solution set is {1.41}.

13\.

$$2e^x = 12.4$$
$$e^x = 6.2$$
$$\ln e^x = \ln 6.2$$
$$x \ln e = \ln 6.2$$
$$x = \ln 6.2 = 1.82$$

The solution set is {1.82}.

17.
$$\begin{aligned} 5^{x-1} &= 2^{2x+1} \\ \ln 5^{x-1} &= \ln 2^{2x+1} \\ (x-1)\ln 5 &= (2x+1)\ln 2 \\ x \ln 5 - \ln 5 &= 2x \ln 2 + \ln 2 \\ x \ln 5 - 2x \ln 2 &= \ln 2 + \ln 5 \\ x(\ln 5 - 2 \ln 2) &= \ln 2 + \ln 5 \\ x &= \frac{\ln 2 + \ln 5}{\ln 5 - 2\ln 2} \\ x &= \frac{\ln 2 + \ln 5}{\ln 5 - \ln 4} \qquad (2 \ln 2 = \ln 2^2 = \ln 4) \\ x &= 10.32 \end{aligned}$$

The solution set is $\{10.32\}$.

21.
$$\begin{aligned} \log(2x-1) - \log(x-3) &= 1 \\ \log(2x-1) - \log(x-3) &= \log 10 \qquad (1 = \log 10) \\ \log(\frac{2x-1}{x-3}) &= \log 10 \\ \frac{2x-1}{x-3} &= 10 \qquad \text{Apply Property 5.8.} \\ 2x-1 &= 10(x-3) \\ 2x-1 &= 10x-30 \\ 29 &= 8x \\ \frac{29}{8} &= x \end{aligned}$$

The solution set is $\{\frac{29}{8}\}$.

25.
$$\begin{aligned} \log(x+1) - \log(x+2) &= \log \frac{1}{x} \\ \log(\frac{x+1}{x+2}) &= \log \frac{1}{x} \\ \frac{x+1}{x+2} &= \frac{1}{x} \qquad \text{Apply Property 5.8.} \\ x(x+1) &= x+2 \\ x^2+x &= x+2 \\ x^2 &= 2 \\ x &= \pm\sqrt{2} \end{aligned}$$

The solution of $-\sqrt{2}$ must be discarded because $(x+1)$ would be negative and logarithms of negative numbers are not defined. Thus, the solution set is $\{\sqrt{2}\}$.

29. $\log(x^2) = (\log x)^2$

$$2 \log x = (\log x)^2$$
$$0 = (\log x)^2 - 2 \log x$$
$$0 = \log x(\log x - 2)$$
$$\log x = 0 \text{ or } \log x - 2 = 0$$
$$\log x = 0 \text{ or } \log x = 2$$
$$x = 1 \text{ or } x = 100$$

The solution set is $\{1, 100\}$.

33. Let $x = \log_5 2.1$. Then by changing to exponential form, we obtain $5^x = 2.1$. This equation can be solved as follows.

$$5^x = 2.1$$
$$\ell n\ 5^x = \ell n\ 2.1$$
$$x\ \ell n\ 5 = \ell n\ 2.1$$
$$x = \frac{\ell n\ 2.1}{\ell n\ 5} = .461$$

37. Let $x = \log_9 14.32$. Then by changing to exponential form, we obtain $9^x = 14.32$. This equation can be solved as follows.

$$9^x = 14.32$$
$$\ell n\ 9^x = \ell n\ 14.32$$
$$x\ \ell n\ 9 = \ell n\ 14.32$$
$$x = \frac{\ell n\ 14.32}{\ell n\ 9} = 1.211$$

41. Using $A = Pe^{rt}$, we obtain

$$1500 = 500e^{.09t}$$
$$3 = e^{.09t}$$
$$\ell n\ 3 = \ell n\ e^{.09t}$$
$$\ell n\ 3 = .09t(\ell n\ e)$$
$$\ell n\ 3 = .09t$$
$$\frac{\ell n\ 3}{.09} = t$$
$$12.2 = t$$

It will take approximately 12.2 years.

45. $15000 = 30000(.9)^5$

$$.5 = (.9)^t$$
$$\ell n(.5) = \ell n(.9)^t$$
$$\ell n(.5) = t\ \ell n(.9)$$
$$\frac{\ell n(.5)}{\ell n(.9)} = t$$
$$6.6 = t$$

It will take approximately 6.6 years.

49. $100000 = 50000e^{.02t}$

$$2 = e^{.02t}$$
$$\ell n\ 2 = \ell n\ e^{.02t}$$
$$\ell n\ 2 = (.02t)\,\ell n\ e$$
$$\ell n\ 2 = .02t \qquad (\ell n\ e = 1)$$
$$\frac{\ell n\ 2}{.02} = t$$
$$34.7 = t$$

It would take approximately 34.7 years.

53. $$\frac{(10^{7.3})(I_0)}{(10^{6.4})(I_0)} = 10^{7.3-6.4} = 10^{.9} \approx 7.9$$

Thus, an earthquake with a Richter number of 7.3 is approximately 8 times as intense as an earthquake with a Richter number of 6.4.

CHAPTER 6

Problem Set 6.1

1.
$$\begin{array}{r|l} & 4x+5 \\ \hline 3x-2 & 12x^2+7x-10 \\ & \underline{12x^2-8x} \\ & \qquad 15x-10 \\ & \qquad \underline{15x-10} \end{array}$$

5.
$$\begin{array}{r|l} & 2x+5 \\ \hline 3x^2+2 & 6x^2+19x+11 \\ & \underline{6x^2+\ 4x} \\ & \qquad 15x+11 \\ & \qquad \underline{15x+10} \\ & \qquad\qquad 1 \end{array}$$

9.
$$\begin{array}{r|l} & 5y-1 \\ \hline y^2-y & 5y^3-6y^2-7y-2 \\ & \underline{5y^3-5y^2} \\ & \quad -y^2-7y-2 \\ & \quad \underline{-y^2+\ y} \\ & \qquad -8y-2 \end{array}$$

13.
$$\begin{array}{r|l} & 3x+4y \\ \hline x-2y & 3x^2-2xy-8y^2 \\ & \underline{3x^2-6xy} \\ & \qquad 4xy-8y^2 \\ & \qquad \underline{4xy-8y^2} \end{array}$$

17.
$$\begin{array}{r|rrr} 4 & 1 & 2 & -10 \\ & & 4 & 24 \\ \hline & 1 & 6 & 14 \end{array}$$
Q: $x+6$ R: 14

21.
$$\begin{array}{r|rrrr} 2 & 1 & -2 & -1 & 2 \\ & & 2 & 0 & -2 \\ \hline & 1 & 0 & -1 & 0 \end{array}$$
Q: x^2-1 R: 0

25.
$$\begin{array}{r|rrrr} -2 & 1 & 0 & -7 & -6 \\ & & -2 & 4 & 6 \\ \hline & 1 & -2 & -3 & 0 \end{array}$$
Q: x^2-2x-3 R: 0

29.
$$\begin{array}{r|rrrr} -3 & 1 & 6 & 11 & 6 \\ & & -3 & -9 & -6 \\ \hline & 1 & 3 & 2 & 0 \end{array}$$
Q: x^2+3x+2 R: 0

33.
$$\begin{array}{r|rrrrrr} 1 & 1 & 0 & 0 & 0 & 0 & 1 \\ & & 1 & 1 & 1 & 1 & 1 \\ \hline & 1 & 1 & 1 & 1 & 1 & 2 \end{array}$$
Q: $x^4+x^3+x^2+x+1$ R: 2

37.
$$\begin{array}{r|rrrrr} \frac{1}{2} & 4 & 0 & -5 & 0 & 1 \\ & & 2 & 1 & -2 & -1 \\ \hline & 4 & 2 & -4 & -2 & 0 \end{array}$$
Q: $4x^3+2x^2-4x-2$ R: 0

Problem Set 6.2

1. (a)
$$\begin{array}{r|rrr} 2 & 1 & 1 & -8 \\ & & 2 & 6 \\ \hline & 1 & 3 & \boxed{-2} \end{array}$$
$f(2) = -2$

(b)
$$\begin{aligned} f(x) &= x^2+x-8 \\ f(2) &= 2^2+2-8 \\ &= 4+2-8 \\ &= -2 \end{aligned}$$

5. (a)
$$\begin{array}{r|rrrrr} -2 & 1 & -2 & -3 & 5 & -1 \\ & & -2 & 8 & -10 & 10 \\ \hline & 1 & -4 & 5 & -5 & \boxed{9} \end{array}$$
$f(-2) = 9$

(b)
$$\begin{aligned} f(x) &= x^4-2x^3-3x^2+5x-1 \\ f(-2) &= (-2)^4-2(-2)^3-3(-2)^2+5(-2)-1 \\ &= 16+16-12-10-1 \\ &= 32-23 \\ &= 9 \end{aligned}$$

9. (a)

$$\begin{array}{r|rrrrr} 3 & 3 & -2 & 0 & 4 & -1 \\ & & 9 & 21 & 63 & 201 \\ \hline & 3 & 7 & 21 & 67 & \boxed{200} \end{array}$$

$f(3) = 200$

(b)

$$\begin{aligned} f(n) &= 3n^4-2n^3+4n-1 \\ f(3) &= 3(3)^4-2(3)^3+4(3)-1 \\ &= 243-54+12-1 \\ &= 200 \end{aligned}$$

13.

$$\begin{array}{r|rrrrr} 7 & 1 & -8 & 9 & -15 & 2 \\ & & 7 & -7 & 14 & -7 \\ \hline & 1 & -1 & 2 & -1 & \boxed{-5} \end{array} \longrightarrow f(7) = -5$$

17.

$$\begin{array}{r|rrrr} \frac{1}{2} & 2 & -5 & 4 & -3 \\ & & 1 & -2 & 1 \\ \hline & 2 & -4 & 2 & \boxed{-2} \end{array} \longrightarrow f(\tfrac{1}{2}) = -2$$

21.

$$\begin{array}{r|rrrr} -2 & 1 & 1 & -7 & -10 \\ & & -2 & 2 & 10 \\ \hline & 1 & -1 & -5 & \boxed{0} \end{array} \longleftarrow$$ The remainder is 0; therefore, $x+2$ is a factor.

25.

$$\begin{array}{r|rrrr} 2 & 1 & 0 & 0 & -8 \\ & & 2 & 4 & 8 \\ \hline & 1 & 2 & 4 & \boxed{0} \end{array} \longleftarrow$$ The remainder is 0; therefore, $x-2$ is a factor.

29.

$$\begin{array}{r|rrrr} -2 & 1 & 7 & 4 & -12 \\ & & -2 & -10 & 12 \\ \hline & 1 & 5 & -6 & 0 \end{array}$$

$$\begin{aligned} x^3+7x^2+4x-12 &= (x+2)(x^2+5x-6) \\ &= (x+2)(x+6)(x-1) \end{aligned}$$

33.

$$\begin{array}{r|rrrr} -1 & 1 & -2 & -7 & -4 \\ & & -1 & 3 & 4 \\ \hline & 1 & -3 & -4 & 0 \end{array}$$

$$\begin{aligned} x^3-2x^2-7x-4 &= (x+1)(x^2-3x-4) \\ &= (x+1)(x-4)(x+1) \\ &= (x+1)^2(x-4) \end{aligned}$$

37. $f(-2)$ must equal zero. Therefore,

$$\begin{aligned} (-2)^3+4(-2)^2-11(-2)+k &= 0 \\ -8 + 16 + 22 + k &= 0 \\ 30+k &= 0 \\ k &= -30 \end{aligned}$$

41. Let $f(x) = x^n-1$. Therefore, $f(1) = 1^n-1 = 1-1 = 0$ for all positive integral values of n. Thus, $x-1$ is a factor of x^n-1.

45. (a)

$$\begin{array}{r|rrr} 1+i & 1 & 4 & -2 \\ & & 1+i & 4+6i \\ \hline & 1 & 5+i & \boxed{2+6i} \end{array} \longrightarrow f(1+i) = 2+6i$$

(b)

$$\begin{aligned} f(x) &= x^2+4x-2 \\ f(1+i) &= (1+i)^2+4(1+i)-2 \\ &= 2i+4+4i-2 \\ &= 2+6i \end{aligned}$$

49. (a) $f(x) = x^3+5x^2-2x+1 = x(x^2+5x-2)+1$
$= x[x(x+5)-2]+1$

$f(4) = 4[4(4+5)-2]+1 = 136+1 = 137$

$f(-5) = -5[-5(-5+5)-2]+1 = 10+1 = 11$

$f(7) = 7[7(7+5)-2]+1 = 574+1 = 575$

Problem Set 6.3

For Problems 1,5,9,13,17,21, and 25 we will use the Rational Root Theorem to determine the possible rational solutions. If $\frac{c}{d}$ is a rational solution of $a_nx^n+\ldots+a_1x+a_0 = 0$, then c is a factor of the constant term a_0 and d is a factor of the leading coefficient a_n.

1. $\frac{c}{d}$: ±1, ±2, ±4

-1	1	1	-4	-4
		-1	0	4
	1	0	-4	0

$x^3+x^2-4x-4 = 0$

$(x+1)(x^2-4) = 0$

$(x+1)(x+2)(x-2) = 0$

$x+1 = 0$ or $x+2 = 0$ or $x-2 = 0$

$x = -1$ or $x = -2$ or $x = 2$

The solution set is $\{-2,-1,2\}$.

5. c: ±1, ±2, ±4, ±7, ±28

d: ±1, ±3

$\frac{c}{d}$: $\pm1, \pm\frac{1}{3}, \pm2, \pm\frac{2}{3}, \pm4, \pm\frac{4}{3}, \pm7, \pm\frac{7}{3}, \pm28, \pm\frac{28}{3}$

2	3	13	-52	28
		6	38	-28
	3	19	-14	0

$3x^3+13x^2-52x+28 = 0$

$(x-2)(3x^2+19x-14) = 0$

$(x-2)(3x-2)(x+7) = 0$

$x-2 = 0$ or $3x-2 = 0$ or $x+7 = 0$

$x = 2$ or $x = \frac{2}{3}$ or $x = -7$

The solution set is $\{-7,\frac{2}{3},2\}$.

9. $\frac{c}{d}$: ±1, ±2, ±3, ±4, ±6, ±8, ±12, ±24

1	1	-4	-7	34	-24
		1	-3	-10	24
2	1	-3	-10	24	0
		2	-2	-24	
	1	-1	-12	0	

$x^4-4x^3-7x^2+34x-24 = 0$

$(x-1)(x-2)(x^2-x-12) = 0$

$(x-1)(x-2)(x-4)(x+3) = 0$

$x-1 = 0$ or $x-2 = 0$ or $x-4 = 0$ or $x+3 = 0$

$x = 1$ or $x = 2$ or $x = 4$ or $x = -3$

The solution set is $\{-3,1,2,4\}$.

13. c: $\pm 1, \pm 2, \pm 4$

d: $\pm 1, \pm 3$

$\frac{c}{d}$: $\pm 1, \pm\frac{1}{3}, \pm 2, \pm\frac{2}{3}, \pm 4, \pm\frac{4}{3}$

$$\begin{array}{r|rrrrr} 1 & 3 & -1 & -8 & 2 & 4 \\ & & 3 & 2 & -6 & -4 \\ \hline -\frac{2}{3} & 3 & 2 & -6 & -4 & 0 \\ & & -2 & 0 & 4 & \\ \hline & 3 & 0 & -6 & 0 & \end{array}$$

$$3x^4-x^3-8x^2+2x+4 = 0$$

$$(x-1)(x+\tfrac{2}{3})(3x^2-6) = 0$$

$$x-1 = 0 \text{ or } x+\tfrac{2}{3} = 0 \text{ or } 3x^2-6 = 0$$

$$x = 1 \text{ or } \quad x = -\tfrac{2}{3} \text{ or } \quad x = \pm\sqrt{2}$$

The solution set is $\{-\frac{2}{3}, 1, \pm\sqrt{2}\}$.

17. $\frac{c}{d}$: $\pm 1, \pm 2, \pm 4$

$$\begin{array}{r|rrrrr} -1 & 1 & -3 & 2 & 2 & -4 \\ & & -1 & 4 & -6 & 4 \\ \hline 2 & 1 & -4 & 6 & -4 & 0 \\ & & 2 & -4 & 4 & \\ \hline & 1 & -2 & 2 & 0 & \end{array}$$

$$x^4-3x^3+2x^2+2x-4 = 0$$

$$(x+1)(x-2)(x^2-2x+2) = 0$$

$$x+1 = 0 \text{ or } x-2 = 0 \text{ or } x^2-2x+2 = 0$$

$$x = -1 \text{ or } \quad x = 2 \text{ or } \quad x = \frac{2 \pm\sqrt{4-8}}{2}$$

$$x = \frac{2 \pm\sqrt{-4}}{2}$$

$$x = \frac{2 \pm 2i}{2}$$

$$x = 1+i$$

The solution set is $\{-1, 2, 1 \pm i\}$.

21. $\frac{c}{d}$: ± 1

$$\begin{array}{r|rrrrr} 1 & 1 & -1 & -8 & -3 & 1 \\ & & 1 & 0 & -8 & -11 \\ \hline & 1 & 0 & -8 & -11 & -10 \end{array} \longleftarrow \text{1 is not a solution.}$$

$$\begin{array}{r|rrrrr} -1 & 1 & -1 & -8 & -3 & 1 \\ & & -1 & 2 & 6 & -3 \\ \hline & 1 & -2 & -6 & 3 & -2 \end{array} \longleftarrow \text{-1 is not a solution.}$$

Since -1 and 1 are the only possible rational solutions and neither of them is a solution, there are no rational solutions.

25. $\frac{c}{d}$: ± 1

$$\begin{array}{r|rrrrrr} 1 & 1 & -2 & 3 & 4 & 7 & -1 \\ & & 1 & -1 & 2 & 6 & 13 \\ \hline & 1 & -1 & 2 & 6 & 13 & 12 \end{array} \longleftarrow \text{ 1 is not a solution.}$$

$$\begin{array}{r|rrrrrr} -1 & 1 & -2 & 3 & 4 & 7 & -1 \\ & & -1 & 3 & -6 & 2 & -9 \\ \hline & 1 & -3 & 6 & -2 & 9 & -10 \end{array} \longleftarrow \text{ -1 is not a solution.}$$

Neither 1 nor -1 is a solution and they are the only rational possibilities.

29. There are 2 variations of sign in $8x^2-14x+3$. Therefore, the given equation has 2 or 0 positive solutions.

Replacing x with -x in the given polynomial produces $8(-x)^2-14(-x)+3$ which simplifies to $8x^2+14x+3$. There are no variations of signs in $8x^2+14x+3$. Therefore, the given equation does not have any negative solutions.

Therefore, the given equation must have either 2 positive solutions or 2 nonreal complex solutions.

33. There is 1 sign variation in $4x^3+5x^2-6x-2$. So, the given equation must have 1 positive solution.

Replacing x with -x in the given polynomial produces $4(-x)^3+5(-x)^2-6(-x)-2$ which simplifies to $-4x^3+5x^2+6x-2$. There are 2 sign variations in $-4x^3+5x^2+6x-2$. Thus, the given equation has 2 or 0 negative solutions.

Therefore, the given equation has 1 positive and 2 negative solutions, or 1 positive and 2 nonreal complex solutions.

37. There is 1 sign variation in $2x^6+3x^4-2x^2-1$. Thus, the given equation has 1 positive solution.

Replacing x with -x produces $2(-x)^6+3(-x)^4-2(-x)^2-1$ which simplifies to $2x^6+3x^4-2x^2-1$ and this polynomial has 1 sign variation. Therefore, the given equation has 1 negative solution.

The given equation must have 1 positive, 1 negative, and 4 nonreal complex solutions.

41. If 3, $-\frac{2}{3}$, and $\frac{3}{4}$ are to be roots, then $(x-3)$, $(x+\frac{2}{3})$, and $(x-\frac{3}{4})$ are to be factors of the polynomial.

$$(x-3)(x+\frac{2}{3})(x-\frac{3}{4}) = 0$$

Now we can multiply both sides of the equation by 12 where we multiply the factor $(x+\frac{2}{3})$ by 3 and the factor $(x-\frac{3}{4})$ by 4.

$$(x-3)(3x+2)(4x-3) = 0$$

Now we can perform the multiplications on the left side.

$$(x-3)(12x^2-x-6) = 0$$

$$12x^3-x^2-6x-36x^2+3x+18 = 0$$

$$12x^3-37x^2-3x+18 = 0$$

45. If 1-4i is a solution, then so is its conjugate, 1+4i.

$$(x+2)[x-(1-4i)][x-(1+4i)] = 0$$
$$(x+2)[(x-1)+4i][(x-1)-4i] = 0$$
$$(x+2)[(x-1)^2-16i^2] = 0$$
$$(x+2)[x^2-2x+1+16] = 0$$
$$(x+2)(x^2-2x+17) = 0$$
$$x^3-2x^2+17x+2x^2-4x+34 = 0$$
$$x^3+13x+34 = 0$$

Problem Set 6.4

1. This is the graph of $f(x) = x^3$ moved down 3 units.

5. This is the graph of $f(x) = x^4$ moved down 2 units.

9. This is the graph of $f(x) = x^5$ moved 1 unit to the right and 2 units up.

13. First, find the x-intercepts by setting each factor equal to zero and solving for x.

$$x+4 = 0 \text{ or } x+1 = 0 \text{ or } 1-x = 0$$
$$x = -4 \text{ or } x = -1 \text{ or } 1 = x$$

Thus, the points (-4,0),(-1,0), and (1,0) are on the graph. These points divide the x-axis into four intervals and the following chart uses a test value in each of those intervals to determine the sign behavior of the function.

Interval	Test value	Sign of f(x)	Location of graph
$x < -4$	$f(-5) = 24$	Positive	Above x-axis
$-4 < x < -1$	$f(-3) = -8$	Negative	Below x-axis
$-1 < x < 1$	$f(0) = 4$	Positive	Above x-axis
$x > 1$	$f(2) = -18$	Negative	Below x-axis

The points (-5,24),(-4,0),(-3,-8),(-1,0),(0,4),(1,0), and (2,-18) along with the information from the chart can be used to sketch the curve.

17. First, find the x-intercepts by setting each factor equal to zero and solving for x.

$$x+3 = 0 \text{ or } x+1 = 0 \text{ or } x-1 = 0 \text{ or } x-2 = 0$$
$$x = -3 \text{ or } x = -1 \text{ or } x = 1 \text{ or } x = 2$$

Thus, the points (-3,0),(-1,0),(1,0), and (2,0) are on the graph. These points also divide the x-axis into five intervals and the following chart uses a test value from each interval to determine the sign behavior of the function.

Interval	Test value	Sign of f(x)	Location of graph
$x < -3$	$f(-4) = 90$	Positive	Above x-axis
$-3 < x < -1$	$f(-2) = -12$	Negative	Below x-axis
$-1 < x < 1$	$f(0) = 6$	Positive	Above x-axis
$1 < x < 2$	$f(\frac{3}{2}) = -\frac{45}{16}$	Negative	Below x-axis
$x > 3$	$f(3) = 48$	Positive	Above x-axis

The points (-4,90),(-3,0),(-2,-12),(-1,0),(0,6),(1,0),$(\frac{3}{2}, -\frac{45}{16})$,(2,0), and (3,48) can be used to sketch the curve.

21. First, find the x-intercepts.

$$x+1 = 0 \text{ or } x-1 = 0$$
$$x = -1 \text{ or } x = 1$$

Thus, the points (-1,0) and (1,0) are on the graph. They also divide the x-axis into 3 intervals and the following chart uses a test value from each interval to determine the sign behavior of the function.

Interval	Test value	Sign of f(x)	Location of graph
$x < -1$	$f(-2) = 9$	Positive	Above x-axis
$-1 < x < 1$	$f(0) = 1$	Positive	Above x-axis
$x > 1$	$f(2) = 9$	Positive	Above x-axis

The points (-2,9),(-1,0),(0,1),(1,0), and (2,9) along with the information from the chart can be used to sketch the curve.

25. $f(x) = -x^4-3x^3-2x^2$

$$= -x^2(x^2+3x+2)$$
$$= -x^2(x+2)(x+1)$$

Now we can find the x-intercepts by setting each factor equal to zero and solving for x.

$$-x^2 = 0 \text{ or } x+2 = 0 \text{ or } x+1 = 0$$
$$x = 0 \text{ or } x = -2 \text{ or } x = -1$$

The points (-2,0),(-1,0), and (0,0) are on the graph and divide the x-axis into four intervals. The following chart uses a test value from each interval to determine the sign behavior of the given function in that interval.

Interval	Test value	Sign of f(x)	Location of graph
$x < -2$	$f(-3) = -18$	Negative	Below x-axis
$-2 < x < -1$	$f(-\frac{3}{2}) = \frac{9}{16}$	Positive	Above x-axis
$-1 < x < 0$	$f(-\frac{1}{2}) = -\frac{3}{16}$	Negative	Below x-axis
$x > 0$	$f(1) = -6$	Negative	Below x-axis

The points (-3,-18),(-2,0),$(-\frac{3}{2}, \frac{9}{16})$,(-1,0),$(-\frac{1}{2}, -\frac{3}{16})$,(0,0), and (1,-6) determine the curve.

29. First, let's use the Rational Root Theorem.

$\frac{c}{d}$: ±1, ±2, ±3, ±4, ±6, ±12

$$\begin{array}{r|rrrr} 1 & 1 & 0 & -13 & 12 \\ & & 1 & 1 & -12 \\ \hline & 1 & 1 & -12 & 0 \end{array}$$

$f(x) = x^3-13x+12$

$= (x-1)(x^2+x-12)$

$= (x-1)(x+4)(x-3)$

Now you can proceed as in the previous problems.

33. First, let's use the Rational Root Theorem.

$\frac{c}{d}$: ±1, ±2, ±3, ±6

$$\begin{array}{r|rrrr} 1 & -1 & 6 & -11 & 6 \\ & & -1 & 5 & -6 \\ \hline & -1 & 5 & -6 & 0 \end{array}$$

$f(x) = -x^3+6x^2-11x+6$

$= (x-1)(-x^2+5x-6)$

$= (x-1)(x-3)(-x+2)$

$= (x-1)(x-3)(2-x)$

Now you can proceed as in the previous problems.

37. (a) $f(0) = (0-4)^2(0+3)^3 = (-4)^2(3)^3 = 432$

(b) Set each factor equal to zero and solve for x.

$(x-4)^2 = 0$ or $(x+3)^3 = 0$
$x = 4$ or $x = -3$

(c) The points (4,0) and (-3,0) divide the x-axis into three intervals. A test value from each interval can be used to determine the sign behavior of the function in that interval.

$f(-4) = (-8)^2(-1)^3 = -64$ Negative

$f(0) = (-4)^2(3)^3 = 432$ Positive

$f(5) = (1)^2(8)^3 = 512$ Positive

Therefore, $f(x) > 0$ if x is in the intervals $(-3,4)$ or $(4,\infty)$ and $f(x) < 0$ if x is in the interval $(-\infty,-3)$.

41. (a) $f(0) = (0+2)^5(0-4)^2 = (2)^5(-4)^2 = 512$

(b) Set each factor equal to zero and solve for x.

$(x+2)^5 = 0$ or $(x-4)^2 = 0$
$x = -2$ or $x = 4$

(c) The points $(-2,0)$ and $(4,0)$ divide the x-axis into three intervals. A test value from each interval can be used to determine the sign behavior of the function in that interval.

$f(-3) = (-1)^5(-7)^2 = -49$ Negative

$f(0) = (2)^5(-4)^2 = 512$ Positive

$f(5) = (7)^5(1)^2 = 16,807$ Positive

Therefore, $f(x) > 0$ if x is in the intervals $(-2,4)$ or $(4,\infty)$, and $f(x) < 0$ if x is in the interval $(-\infty,-2)$.

Problem Set 6.5

1. First, notice that $f(-x) = \frac{-1}{-x} = \frac{1}{x}$ and this equals the opposite of f(x). Therefore, this graph has origin symmetry.

   Secondly, the line $x = 0$ (vertical axis) is a vertical asymptote.

   Thirdly, by letting x take on larger and larger values, we see that f(x) approaches zero from below. So, $f(x) = 0$ (the x-axis) is a horizontal asymptote.

   Finally, by plotting some points in the fourth quadrant (use positive values for x), one branch of the curve can be drawn. Then, by origin symmetry we can draw the other branch in the second quadrant.

5. The line $x = 1$ is a vertical asymptote.

   As x gets larger and larger, the value of f(x) approaches zero from above. Therefore, the line $f(x) = 0$ is a horizontal asymptote.

   Plotting some points on both sides of the vertical asymptote should determine the two branches of the curve.

9. The line $x = -2$ is a vertical asymptote.

   Divide both numerator and denominator of the given function by x.

$$f(x) = \frac{-3x}{x+2} = \frac{-3}{1 + \frac{2}{x}}$$

   Now we can see that as x gets larger and larger, the value of f(x) approaches -3 from above. Thus, the line $f(x) = -3$ is a horizontal asymptote.

   Finally, by plotting some points on both sides of the vertical asymptote we can determine the two branches of the curve.

13. The lines $x = -1$ and $x = 2$ are vertical asymptotes.

   By letting x get larger and larger, we see that f(x) approaches zero from below. Therefore, $f(x) = 0$ is a horizontal asymptote.

   By plotting points in the three regions determined by the two vertical asymptotes, we can determine the three branches of the curve.

17. The line $x = 0$ is a vertical asymptote.

Divide both numerator and denominator of the given function by x.

$$f(x) = \frac{x+2}{x} = \frac{1 + \frac{2}{x}}{1}$$

Now we can see that as x gets larger and larger, the value of f(x) approaches 1 from above. Therefore, the line $f(x) = 1$ is a horizontal asymptote.

Finally, by plotting points on both sides of the vertical asymptote, we can determine the two branches of the curve.

21. Since $f(-x) = \frac{2(-x)^4}{(-x)^4+1} = \frac{2x^4}{x^4+1} = f(x)$, the curve is symmetric with respect to the vertical axis.

The denominator, x^4+1, cannot equal zero; thus, there is no vertical asymptote.

Divide both numerator and denominator of the given function by x^4.

$$f(x) = \frac{2x^4}{x^4+1} = \frac{2}{1+\frac{1}{x^4}}.$$

Now we can see that as x gets larger and larger, the value of f(x) approaches 2 from below. Therefore, the line $f(x) = 2$ is a horizontal asymptote.

The right half of the curve can be determined using positive values for x. Then, by symmetry, we can sketch the left half of the curve.

Problem Set 6.6

The suggestions for the problems in this problem set are given in a step-by-step format. The order of the steps is flexible, but this indicates the order in which I usually do such problems.

1. (a) Identify the vertical asymptotes by setting the denominator equal to zero and solving for x.

$$x^2+x-2 = 0$$
$$(x+2)(x-1) = 0$$
$$x+2 = 0 \text{ or } x-1 = 0$$
$$x = -2 \text{ or } x = 1$$

So, the lines $x = -2$ and $x = 1$ are vertical asymptotes.

(b) Divide both numerator and denominator of given function by x^2.

$$f(x) = \frac{x^2}{x^2+x-2} = \frac{1}{1 + \frac{1}{x} - \frac{2}{x^2}}$$

Now we can see that as x gets larger and larger, the value of f(x) approaches 1 from below. Thus, the line $f(x) = 1$ is a horizontal asymptote.

(c) To determine if any points of the graph are on the horizontal asymptote we can see if the equation

$$\frac{x}{x^2+x-2} = 1$$

has any solutions.

$$\frac{x^2}{x^2+x-2} = 1$$

$$x^2 = x^2+x-2$$
$$0 = x-2$$
$$2 = x$$

Therefore, the point (2,1) is on the graph.

(d) Now by plotting some points in the three regions determined by the two vertical asymptotes, the three branches of the curve can be determined.

5. (a) Notice that $f(-x) = \frac{x}{x^2-1} = -f(x)$. Thus, the curve is symmetric with respect to the origin.

(b) Find the vertical asymptotes by setting the denominator equal to zero and solving for x.

$$x^2-1 = 0$$
$$x^2 = 1$$
$$x = \pm 1$$

Thus, the lines x = 1 and x = -1 are vertical asymptotes.

(c) Divide both numerator and denominator of the given function by x^2.

$$f(x) = \frac{-x}{x^2-1} = \frac{-\frac{1}{x}}{1 - \frac{1}{x^2}}$$

Now we can see that as x gets larger and larger, the value of f(x) approaches zero from below. Thus, the line f(x) = 0 is a horizontal asymptote.

(d) Set the given function equal to zero and solve for x to see if any points of the graph are on the horizontal asymptote.

$$\frac{-x}{x^2-1} = 0$$

$$-x = 0$$

$$x = 0$$

The origin, (0,0), is a point of the graph.

(e) Now plot points on both sides of x = 1, using positive values for x. This will determine the portion of the curve in the first and fourth quadrants. Then reflect this portion through the origin into the second and third quadrants.

9. (a) Find the vertical asymptotes by setting the denominator equal to zero and solving for x.

$$x^2-4x+3 = 0$$
$$(x-3)(x-1) = 0$$
$$x-3 = 0 \text{ or } x-1 = 0$$
$$x = 3 \text{ or } \quad x = 1$$

The lines $x = 3$ and $x = 1$ are vertical asymptotes.

(b) Divide both numerator and denominator of the given function by x^2.

$$f(x) = \frac{x^2}{x^2-4x+3} = \frac{1}{1 - \frac{4}{x} + \frac{3}{x^2}}$$

Now we can see that as x gets larger and larger, the value of f(x) approaches 1 from above. Thus, the line $f(x) = 1$ is a horizontal asymptote.

(c) Set the given function equal to 1 to see if any points of the graph are on the horizontal asymptote.

$$\frac{x^2}{x^2-4x+3} = 1$$

$$x^2 = x^2-4x+3$$

$$0 = -4x+3$$

$$4x = 3$$

$$x = \frac{3}{4}$$

Thus, the point $(\frac{3}{4},1)$ is a point of the graph.

(d) Now by plotting some points in each of the three regions determined by the two vertical asymptotes, you should be able to determine the curve.

13. (a) First, notice that $f(-x) = \frac{4x}{x^2+1} = -f(x)$. Therefore, this curve is symmetric with respect to the origin.

(b) Since x^2+1 cannot equal zero for any real values of x, there are no vertical asymptotes.

(c) Divide both numerator and denominator of given function by x^2.

$$f(x) = \frac{-4x}{x^2+1} = \frac{\frac{-4}{x}}{1 + \frac{1}{x^2}}$$

Now we can see that as x gets larger and larger, the value of f(x) approaches 0 from below. Thus, the line $f(x) = 0$ (the x-axis) is a horizontal asymptote.

(d) Since $f(0) = 0$, the point (0,0) is a point of the graph.

(e) Now plot some points in the fourth quadrant using positive values for x. Then these points determine the part of the curve located in the fourth quadrant. Finally, by reflecting this portion of the curve through the origin, the complete curve can be drawn.

17. (a) Since x+1 equals zero when $x = -1$, the line $x = -1$ is a vertical asymptote.

(b) Since the degree of the numerator is greater than the degree of the denominator, we can change the form of the given rational expression by division.

$$\begin{array}{r|rrr} -1 & 1 & -1 & -6 \\ & & -1 & 2 \\ \hline & 1 & -2 & -4 \end{array}$$

Therefore, the original function can be written as

$$f(x) = \frac{x^2-x-6}{x+1} = x - 2 - \frac{4}{x+1}$$

So the line $f(x) = x-2$ is an oblique asymptote.

(c) Now by plotting a sufficient number of points on both sides of the vertical asymptote, you should be able to sketch the curve.

Problem Set 6.7

1. According to part 1. of Property 6.7, each of the linear factors of the denominator produces a partial fraction of the form "constant over linear factor." In other words, we can write

$$\frac{11x-10}{(x-2)(x+1)} = \frac{A}{x-2} + \frac{B}{x+1} \qquad (1)$$

for some constants A and B. To find A and B, let's multiply both sides of equation (1) by the LCD, $(x-2)(x+1)$, producing

$$11x-10 = A(x+1)+B(x-2). \qquad (2)$$

Equation (2) is an identity; it is true for all values of x. Therefore, let's choose some convenient values for x that will determine the values for A and B.

If we let $x = -1$, then equation (2) becomes

$$\begin{aligned} 11(-1)-10 &= A(-1+1)+B(-1-2) \\ -21 &= -3B \\ 7 &= B. \end{aligned}$$

If we let $x = 2$, then equation (2) becomes

$$\begin{aligned} 11(2)-10 &= A(2+1)+B(2-2) \\ 12 &= 3A \\ 4 &= A. \end{aligned}$$

Therefore, the given rational expression can be written as

$$\frac{11x-10}{(x-2)(x+1)} = \frac{4}{x-2} + \frac{7}{x+1} .$$

5. First, let's factor the denominator.

$$\frac{20x-3}{6x^2+7x-3} = \frac{20x-3}{(2x+3)(3x-1)}$$

Now according to part 1. of Property 6.7, we can write

$$\frac{20x-3}{(2x+3)(3x-1)} = \frac{A}{2x+3} + \frac{B}{3x-1} \,. \qquad (1)$$

Multiplying both sides of (1) by (2x+3)(3x-1) produces

$$20x-3 = A(3x-1)+B(2x+3). \qquad (2)$$

If $x = \frac{1}{3}$, then (2) becomes

$$20(\tfrac{1}{3})-3 = A(3(\tfrac{1}{3})-1)+B(2(\tfrac{1}{3})+3)$$

$$\frac{11}{3} = \frac{11}{3}B$$

$$1 = B.$$

If $x = -\frac{3}{2}$, then (2) becomes

$$20(-\tfrac{3}{2})-3 = A(3(-\tfrac{3}{2})-1) + B(2(-\tfrac{3}{2})+3)$$

$$-33 = -\frac{11}{2}A$$

$$6 = A.$$

Therefore,

$$\frac{20x-3}{6x^2+7x-3} = \frac{6}{2x+3} + \frac{1}{3x-1} \,.$$

9. By part 1. of Property 6.7, we can write

$$\frac{-6x^2+7x+1}{x(2x-1)(4x+1)} = \frac{A}{x} + \frac{B}{2x-1} + \frac{C}{4x+1}. \qquad (1)$$

Mulitplying both sides of (1) by x(2x-1)(4x+1) produces

$$-6x^2+7x+1 = A(2x-1)(4x+1)+B(x)(4x+1)+C(x)(2x-1). \qquad (2)$$

If $x = \frac{1}{2}$, then (2) becomes

$$-6(\tfrac{1}{2})^2 + 7(\tfrac{1}{2})+1 = B(\tfrac{1}{2})(3)$$

$$3 = \frac{3}{2}B$$

$$2 = B.$$

If $x = -\frac{1}{4}$, then (2) becomes

$$-6(-\tfrac{1}{4})^2 + 7(-\tfrac{1}{4})+1 = C(-\tfrac{1}{4})(-\tfrac{3}{2})$$

$$-\frac{9}{8} = \frac{3}{8}C$$

$$-3 = C.$$

If $x = 0$, then (2) becomes

$$-6(0)^2+7(0)+1 = A(-1)(1)$$
$$1 = -A$$
$$-1 = A.$$

Therefore,

$$\frac{-6x^2+7x+1}{x(2x-1)(4x+1)} = \frac{-1}{x} + \frac{2}{2x-1} - \frac{3}{4x+1} .$$

13. Applying parts 1. and 2. of Property 6.7, we can write

$$\frac{-6x^2+19x+21}{x^2(x+3)} = \frac{A}{x} + \frac{B}{x^2} + \frac{C}{x+3} . \qquad (1)$$

Multiplying both sides of (1) by $x^2(x+3)$ produces

$$-6x^2+19x+21 = A(x)(x+3)+B(x+3)+C(x^2). \qquad (2)$$

If $x = 0$, then (2) becomes

$$-6(0)^2+19(0)+21 = 3B$$
$$21 = 3B$$
$$7 = B.$$

If $x = -3$, then (2) becomes

$$-6(-3)^2+19(-3)+21 = 9C$$
$$-90 = 9C$$
$$-10 = C.$$

If $x = 1$, then (2) becomes

$$-6(1)^2+19(1)+21 = 4A+4B+C$$
$$34 = 4A+4B+C.$$

Since we already know that $B = 7$ and $C = -10$, we can determine A.

$$34 = 4A+4(7)+(-10)$$
$$34 = 4A+18$$
$$16 = 4A$$
$$4 = A$$

Therefore,

$$\frac{-6x^2+19x+21}{x^2(x+3)} = \frac{4}{x} + \frac{7}{x^2} - \frac{10}{x+3} .$$

17. According to part 2. of Property 6.7, we can write

$$\frac{3x^2+10x+9}{(x+2)^3} = \frac{A}{x+2} + \frac{B}{(x+2)^2} + \frac{C}{(x+2)^3} . \qquad (1)$$

Multiplying both sides of (1) by $(x+2)^3$ produces

$$3x^2+10x+9 = A(x+2)^2+B(x+2)+C. \qquad (2)$$

If $x = -2$, then (2) becomes

$$3(-2)^2+10(-2)+9 = C$$
$$1 = C.$$

If $x = 0$, then (2) becomes

$9 = 4A+2B+C$; but $C = 1$, so we have

$8 = 4A+2B$ or $4 = 2A+B$.

. If $x = 1$, then (2) becomes

$22 = 9A+3B+C$; but $C = 1$, so we have

$21 = 9A+3B$ or $7 = 3A+B$.

Solving the system $\begin{pmatrix} 2A+B = 4 \\ 3A+B = 7 \end{pmatrix}$ produces $A = 3$ and $B = -2$.

Therefore,

$$\frac{3x^2+10x+9}{(x+2)^3} = \frac{3}{x+2} - \frac{2}{(x+2)^2} + \frac{1}{(x+2)^3}.$$

21. Applying part 4. of Property 6.7, we can write

$$\frac{2x^3+x+3}{(x^2+1)^2} = \frac{Ax+B}{x^2+1} + \frac{Cx+D}{(x^2+1)^2}. \quad (1)$$

Multiplying both sides of (1) by $(x^2+1)^2$ produces

$$2x^3+x+3 = (Ax+B)(x^2+1)+Cx+D. \quad (2)$$

Since equation (2) is an identity, we know that the coefficients of similar terms on both sides of the equation must be equal. Therefore, let's collect similar terms on the right side of equation (2).

$$2x^3+x+3 = Ax^3+Ax+Bx^2+B+Cx+D$$

$$= Ax^3+Bx^2+(A+C)x+B+D$$

Now we can equate coefficients from both sides producing

$2 = A$, $0 = B$, $1 = A+C$, and $3 = B+D$.

From these equations we can determine that $A = 2$, $B = 0$, $C = -1$, and $D = 3$. Therefore, the given expression can be written as

$$\frac{2x^3+x+3}{(x^2+1)^2} = \frac{2x}{x^2+1} + \frac{3-x}{(x^2+1)^2}.$$

CHAPTER 7

Problem Set 7.1

1. The second equation is in appropriate form to begin the substitution process. In the first equation, we can replace y with x+2 and solve for x.

$$x+(x+2) = 16$$
$$2x = 14$$
$$x = 7$$

If x = 7, then using the second equation we obtain

$$y = x+2 = 7+2 = 9.$$

The solution set is $\{(7,9)\}$.

5. The first equation is in appropriate form to begin the substitution process. In the second equation, we can replace y with $\frac{2}{3}x-1$ and then solve for x.

$$5x-7(\frac{2}{3}x-1) = 9$$
$$5x - \frac{14}{3}x + 7 = 9$$
$$15x-14x+21 = 27 \quad \text{Multiply both sides by 3.}$$
$$x+21 = 27$$
$$x = 6$$

If x = 6, then using the first equation, we obtain

$$y = \frac{2}{3}(6)-1 = 3$$

The solution set is $\{(6,3)\}$.

9. The second equation is in appropriate form to begin the substitution process. In the first equation, replace y with $\frac{2}{3}x-\frac{4}{3}$ and then solve for x.

$$2x-3(\frac{2}{3}x-\frac{4}{3}) = 4$$
$$2x-2x+4 = 4$$
$$0 = 0$$

The statement 0 = 0 implies that the system is dependent. That is to say, any ordered pair that satisfies one equation also satisfies the other equation. So if we let x = k, then from the second equation we have $y = \frac{2}{3}k-\frac{4}{3}$. So the solution set of the system can be represented by $\{(k, \frac{2}{3}k - \frac{4}{3})\}$, where k is any real number.

13. Solve the first equation for y in terms of x.

$$4x+3y = -7$$
$$3y = -4x-7$$
$$y = -\frac{4}{3}x-\frac{7}{3}$$

Now substitute $-\frac{4}{3}x-\frac{7}{3}$ for y in the second equation and then solve for x.

$$3x-2(-\frac{4}{3}x-\frac{7}{3}) = 16$$

$$3x+\frac{8}{3}x+\frac{14}{3} = 16$$

$$9x+8x+14 = 48 \quad \text{Multiply both sides by 3.}$$

$$17x = 34$$

$$x = 2$$

Substituting 2 for x in the first equation produces

$$4(2)+3y = -7$$

$$8+3y = -7$$

$$3y = -15$$

$$y = -5$$

The solution set is $\{(2,-5)\}$.

17. Solve the first equation for x in terms of y.

$$4x-5y = 3$$

$$4x = 5y+3$$

$$x = \frac{5}{4}y+\frac{3}{4}$$

Substitute $\frac{5}{4}y+\frac{3}{4}$ for x in the second equation and then solve for y.

$$8(\frac{5}{4}y+\frac{3}{4})+15y = -24$$

$$10y+6+15y = -24$$

$$25y = -30$$

$$y = -\frac{30}{25} = -\frac{6}{5}$$

Substitute $-\frac{6}{5}$ for y in the first equation and solve for x.

$$4x-5(-\frac{6}{5}) = 3$$

$$4x+6 = 3$$

$$4x = -3$$

$$x = -\frac{3}{4}$$

The solution set is $\{(-\frac{3}{4}, -\frac{6}{5})\}$.

21. $\begin{pmatrix} x-3y = -22 \\ 2x+7y = 60 \end{pmatrix}$ (1) (2)

Multiply equation (1) by -2 and add to equation (2) to produce a new equation (4).

$\begin{pmatrix} x-3y = -22 \\ 13y = 104 \end{pmatrix}$ (3) (4)

From equation (4) we see that y = 8. Then substitute 8 for y in equation (3).

$$x-3(8) = -22$$

$$x = 2$$

The solution set is {(2,8)}.

25. $\begin{pmatrix} 5x-2y = 19 \\ 5x-2y = 7 \end{pmatrix}$ (1) (2)

Multiply equation (1) by -1 and add to equation (2) to produce a new equation (4).

$\begin{pmatrix} 5x-2y = 19 \\ 0 = -12 \end{pmatrix}$ (3) (4)

The statement 0 = -12 is obviously false and this implies that the original system is inconsistent. The solution set is ∅.

29. $\begin{pmatrix} \frac{2}{3}s + \frac{1}{4}t = -1 \\ \frac{1}{2}s - \frac{1}{3}t = -7 \end{pmatrix}$ (1) (2)

Multiply equation (1) by 12 and equation (2) by 6.

$\begin{pmatrix} 8s+3t = -12 \\ 3s-2t = -42 \end{pmatrix}$ (3) (4)

Multiply equation (3) by 3 and equation (4) by -8.

$\begin{pmatrix} 24s+9t = -36 \\ -24s+16t = 336 \end{pmatrix}$ (5) (6)

Add equation (5) to equation (6) to produce a new equation (8).

$\begin{pmatrix} 24s+9t = -36 \\ 25t = 300 \end{pmatrix}$ (7) (8)

From equation (8) we see that t = 12. Then substituting 12 for t in equation (7) produces

$$24s+9(12) = -36$$

$$24s = -144$$

$$s = -6$$

So, the solution set is s = -6 and t = 12.

33. $$\begin{pmatrix} \frac{4x}{5} - \frac{3y}{2} = \frac{1}{5} \\ -2x + y = -1 \end{pmatrix} \quad \begin{matrix} (1) \\ (2) \end{matrix}$$

Multiply equation (1) by 10.

$$\begin{pmatrix} 8x-15y = 2 \\ -2x + y = -1 \end{pmatrix} \quad \begin{matrix} (3) \\ (4) \end{matrix}$$

Exchange equations (3) and (4).

$$\begin{pmatrix} -2x + y = -1 \\ 8x-15y = 2 \end{pmatrix} \quad \begin{matrix} (5) \\ (6) \end{matrix}$$

Multiply equation (5) by 4 and add to equation (6) to produce a new equation (8).

$$\begin{pmatrix} -2x + y = -1 \\ -11y = -2 \end{pmatrix} \quad \begin{matrix} (7) \\ (8) \end{matrix}$$

From equation (8) we see that $y = \frac{2}{11}$. Now substitute $\frac{2}{11}$ for y in equation (7).

$$-2x + \frac{2}{11} = -1$$

$$-2x = -\frac{13}{11}$$

$$x = \frac{13}{22}$$

The solution set is $\{(\frac{13}{22}, \frac{2}{11})\}$.

37. Equate the two representations for x from the two equations.

$$3y-10 = -2y+15$$

$$5y = 25$$

$$y = 5$$

Substitute 5 for y in the first equation.

$$x = 3(5)-10 = 5$$

The solution set is $\{(5,5)\}$.

41. $$\begin{pmatrix} \frac{1}{2}x - \frac{2}{3}y = 22 \\ \frac{1}{2}x + \frac{1}{4}y = 0 \end{pmatrix} \quad \begin{matrix} (1) \\ (2) \end{matrix}$$

Multiply equation (1) by 6 and equation (2) by 4.

$$\begin{pmatrix} 3x-4y = 132 \\ 2x + y = 0 \end{pmatrix} \quad \begin{matrix} (3) \\ (4) \end{matrix}$$

Multiply equation (3) by 2 and equation (4) by -3.

$$\begin{pmatrix} 6x-8y = 264 \\ -6x-3y = 0 \end{pmatrix} \quad \begin{matrix} (5) \\ (6) \end{matrix}$$

Add equation (5) to equation (6) to produce a new equation (8).

$$\begin{pmatrix} 6x - 8y = 264 \\ -11y = 264 \end{pmatrix} \quad \begin{matrix} (7) \\ (8) \end{matrix}$$

From equation (8) we can determine that $y = -24$. Then substituting -24 for y in equation (7) produces

$$6x-8(-24) = 264$$
$$6x+192 = 264$$
$$6x = 72$$
$$x = 12$$

The solution set is $\{(12,-24)\}$.

45. $$\begin{pmatrix} x + y = 1000 \\ .12x+.14y = 136 \end{pmatrix} \quad \begin{matrix} (1) \\ (2) \end{matrix}$$

Multiply equation (2) by 100.

$$\begin{pmatrix} x + y = 1000 \\ 12x+14y = 13600 \end{pmatrix} \quad \begin{matrix} (3) \\ (4) \end{matrix}$$

Multiply equation (3) by -12 and add to equation (4) to form a new equation (6).

$$\begin{pmatrix} x + y = 1000 \\ 2y = 1600 \end{pmatrix} \quad \begin{matrix} (5) \\ (6) \end{matrix}$$

From equation (6) we can determine that $y = 800$. Then substituting 800 for y in equation (5) produces

$$x+800 = 1000$$
$$x = 200.$$

The solution set is $\{(200,800)\}$.

49. $$\begin{pmatrix} x + y = 10.5 \\ .5x+.8y = 7.35 \end{pmatrix} \quad \begin{matrix} (1) \\ (2) \end{matrix}$$

Multiply equation (2) by 10.

$$\begin{pmatrix} x + y = 10.5 \\ 5x+8y = 73.5 \end{pmatrix} \quad \begin{matrix} (3) \\ (4) \end{matrix}$$

Multiply equation (3) by -5 and add to equation (4) to produce a new equation (6).

$$\begin{pmatrix} x + y = 10.5 \\ 3y = 21 \end{pmatrix} \quad \begin{matrix} (5) \\ (6) \end{matrix}$$

From equation (6) we can determine that $y = 7$. Then substituting 7 for y in equation (5) produces

$$x+7 = 10.5$$
$$x = 3.5.$$

The solution set is $\{(3.5,7)\}$.

53. Let x and y represent the two numbers. The problem translates into the following system.

$$\begin{pmatrix} y = 3x \\ y-x = 10 \end{pmatrix}$$ $\longleftarrow$ One of the numbers is three times the other. $\longleftarrow$ Their difference is 10.

Because of the format of the first equation, this system lends itself to solving by substitution. Substituting 3x for y in the second equation produces

$$3x-x = 10$$
$$2x = 10$$
$$x = 5.$$

Now substituting 5 for x in the first equation produces

$y = 3x = 3(5) = 15.$

The numbers are 5 and 15.

57. Let u represent the units digit and t the tens digit. The problem translates into the following system.

$$\begin{pmatrix} t+u = 7 \\ 10u+t = 10t+u+9 \end{pmatrix}$$ $\longleftarrow$ The sum of the digits is 7. $\longleftarrow$ The newly formed number is 9 larger than the original number.

Simplify the second equation.

$$\begin{pmatrix} t+u = 7 \\ -t+u = 1 \end{pmatrix}$$

Add the first equation to the second equation to form a new second equation.

$$\begin{pmatrix} t+u = 7 \\ 2u = 8 \end{pmatrix}$$

From $2u = 8$ we see that $u = 4$. Then substitute 4 for u in $t+u = 7$.

$$t+4 = 7$$
$$t = 3$$

The number is 34.

61. Let s represent the number of student tickets and n the number of non-student tickets.

$$\begin{pmatrix} s+n = 3000 \\ 3s+5n = 10000 \end{pmatrix}$$ $\longleftarrow$ 3000 tickets were sold. $\longleftarrow$ Total income was \$10,000.

Multiply the first equation by -3 and add to the second equation to produce a new second equation.

$$\begin{pmatrix} s+n = 3000 \\ 2n = 1000 \end{pmatrix}$$

From $2n = 1000$, we see that $n = 500$. Now substitute 500 for n in $s+n = 3000$.

$$s+500 = 3000$$
$$s = 2500$$

They sold 2500 student tickets and 500 non-student tickets.

65. Let x represent the amount of 40% solution and y the amount of 60% solution to be mixed.

$$\begin{pmatrix} x + y = 20 \\ .4x+.6y = 10.4 \end{pmatrix}$$

$\leftarrow$ total amount of 20 liters of solution

$\leftarrow$ total amount of (52%)(20) = 10.4 liters of pure alcohol

Multiply the second equation by 10.

$$\begin{pmatrix} x + y = 20 \\ 4x+6y = 104 \end{pmatrix}$$

Multiply the first equation by -4 and add to the second equation to form a new second equation.

$$\begin{pmatrix} x+y = 20 \\ 2y = 24 \end{pmatrix}$$

From 2y = 24, we get y = 12. Then substitute 12 for y in x+y = 20.

$$x+12 = 20$$

$$x = 8$$

Therefore, we need to use 8 liters of the 40% solution and 12 liters of the 60% solution.

69. Let x represent the number of five-dollar bills and y the number of ten-dollar bills.

$$\begin{pmatrix} x = y+12 \\ 5x+10y = 330 \end{pmatrix}$$

$\leftarrow$ 12 more five-dollar bills than ten-dollar bills

$\leftarrow$ total amount of $330

Substitute y+12 for x in the second equation.

$$5(y+12)+10y = 330$$

$$5y+60+10y = 330$$

$$15y = 270$$

$$y = 18$$

If y = 18, then x = y+12 = 18+12 = 30.

So, there are 30 five-dollar bills and 18 ten-dollar bills.

73. $$\begin{pmatrix} \frac{3}{x} - \frac{2}{y} = \frac{13}{6} & (1) \\ \frac{2}{x} + \frac{3}{y} = 0 & (2) \end{pmatrix}$$

Multiply (1) by 2 and (2) by -3.

$$\begin{pmatrix} \frac{6}{x} - \frac{4}{y} = \frac{13}{3} & (3) \\ \frac{-6}{x} - \frac{9}{y} = 0 & (4) \end{pmatrix}$$

Add (3) to (4) to produce a new (6).

$$\begin{pmatrix} \frac{6}{x} - \frac{4}{y} = \frac{13}{3} & (5) \\ -\frac{13}{y} = \frac{13}{3} & (6) \end{pmatrix}$$

From (6) we get $y = -3$. Then substitute -3 for y in (5).

$$\frac{6}{x} - \frac{4}{-3} = \frac{13}{3}$$

$$\frac{6}{x} + \frac{4}{3} = \frac{13}{3}$$

$$\frac{6}{x} = \frac{9}{3} = 3$$

$$x = 2$$

The solution set is $\{(2,-3)\}$.

Problem Set 7.2

1. $$\begin{pmatrix} 2x-3y+4z = 10 \\ 5y-2z = -16 \\ 3z = 9 \end{pmatrix} \quad \begin{matrix} (1) \\ (2) \\ (3) \end{matrix}$$

From equation (3) we obtain $z = 3$. Then we can substitute 3 for z in equation (2).

$$5y - 2z = -16$$

$$5y-2(3) = -16$$

$$5y = -10$$

$$y = -2$$

Now we can substitute 3 for z and -2 for y in equation (1).

$$2x - 3y + 4z = 10$$

$$2x-3(-2)+4(3) = 10$$

$$2x + 6 + 12 = 10$$

$$2x = -8$$

$$x = -4$$

The solution set is $\{(-4,-2,3)\}$.

5. $$\begin{pmatrix} 3x+2y-2z = 14 \\ x \quad -6z = 16 \\ 2x \quad +5z = -2 \end{pmatrix} \quad \begin{matrix} (1) \\ (2) \\ (3) \end{matrix}$$

Replace equation (3) with the result of multiplying equation (2) by -2 and adding to equation (3).

$$\begin{pmatrix} 3x+2y-2z = 14 \\ x \quad -6z = 16 \\ 17z = -34 \end{pmatrix} \quad \begin{matrix} (4) \\ (5) \\ (6) \end{matrix}$$

From equation (6) we obtain $z = -2$. Then we can substitute -2 for z in equation (5).

$$x - 6z = 16$$

$$x-6(-2) = 16$$

$$x = 4$$

Now we can substitute 4 for x and -2 for z in equation (4).

$$3x+2y-2z = 14$$
$$3(4)+2y-2(-2) = 14$$
$$12+2y+4 = 14$$
$$2y = -2$$
$$y = -1$$

The solution set is $\{(4,-1,-2)\}$.

9. $$\begin{pmatrix} 2x-y+z = 0 \\ 3x-2y+4z = 11 \\ 5x+y-6z = -32 \end{pmatrix} \begin{matrix} (1) \\ (2) \\ (3) \end{matrix}$$

Replace equation (2) with the result of multiplying equation (1) by -2 and adding to equation (2). Also replace equation (3) with the result of adding equation (1) to equation (3).

$$\begin{pmatrix} 2x-y+z = 0 \\ -x+2z = 11 \\ 7x-5z = -32 \end{pmatrix} \begin{matrix} (4) \\ (5) \\ (6) \end{matrix}$$

Replace equation (6) with the result of multiplying equation (5) by 7 and adding to equation (6).

$$\begin{pmatrix} 2x-y+z = 0 \\ -x+2z = 11 \\ 9z = 45 \end{pmatrix} \begin{matrix} (7) \\ (8) \\ (9) \end{matrix}$$

From equation (9) we obtain $z = 5$. Then we can substitute 5 for z in equation (8).

$$-x+2z = 11$$
$$-x+2(5) = 11$$
$$-x = 1$$
$$x = -1$$

Now we can substitute -1 for x and 5 for z in equation (7).

$$2x-y+z = 0$$
$$2(-1)-y+5 = 0$$
$$-y = -3$$
$$y = 3$$

The solution set is $\{(-1,3,5)\}$.

13. $$\begin{pmatrix} 2x+3y-4z = -10 \\ 4x-5y+3z = 2 \\ 2y+z = 8 \end{pmatrix} \begin{matrix} (1) \\ (2) \\ (3) \end{matrix}$$

Replace equation (2) with the result of multiplying equation (1) by -2 and adding to equation (2).

$$\begin{pmatrix} 2x + 3y - 4z = -10 \\ -11y + 11z = 22 \\ 2y + z = 8 \end{pmatrix} \begin{matrix} (4) \\ (5 \\ (6) \end{matrix}$$

Multiply equation (5) by $\frac{1}{11}$.

$$\begin{pmatrix} 2x+3y-4z = -10 \\ -y + z = 2 \\ 2y + z = 8 \end{pmatrix} \begin{matrix} (7) \\ (8) \\ (9) \end{matrix}$$

Replace equation (9) with the result of multiplying equation (8) by -1 and adding to equation (9).

$$\begin{pmatrix} 2x+3y-4z = -10 \\ -y + z = 2 \\ 3y = 6 \end{pmatrix} \begin{matrix} (10) \\ (11) \\ (12) \end{matrix}$$

From equation (12) we obtain $y = 2$. Then we can substitute 2 for y in equation (11).

$$-y+z = 2$$
$$-2+z = 2$$
$$z = 4$$

Now we can substitute 2 for y and 4 for z in euqation (10).

$$2x + 3y - 4z = -10$$
$$2x+3(2)-4(4) = -10$$
$$2x = 0$$
$$x = 0$$

The solution set is $\{(0,2,4)\}$.

17. $$\begin{pmatrix} 2x-3y+4z = -12 \\ 4x+2y-3z = -13 \\ 6x-5y+7z = -31 \end{pmatrix} \begin{matrix} (1) \\ (2) \\ (3) \end{matrix}$$

Replace equation (2) with the result of multiplying equation (1) by -2 and adding to equation (2). Also replace equation (3) with the result of multiplying equation (1) by -3 and adding to equation (3).

$$\begin{pmatrix} 2x-3y + 4z = -12 \\ 8y-11z = 11 \\ 4y-5z = 5 \end{pmatrix} \begin{matrix} (4) \\ (5) \\ (6) \end{matrix}$$

Replace equation (5) with the result of multiplying equation (6) by -2 and adding to equation (5).

$$\begin{pmatrix} 2x-3y+4z = -12 \\ -z = 1 \\ 4y-5z = 5 \end{pmatrix} \begin{matrix} (7) \\ (8) \\ (9) \end{matrix}$$

From equation (8) we obtain $z = -1$. Then we can substitute -1 for z in equation (9).

$$4y - 5z = 5$$
$$4y-5(-1) = 5$$
$$4y = 0$$
$$y = 0$$

Now we can substitute 0 for y and -1 for z in equation (7).

$$2x - 3y + 4z = -12$$
$$2x-3(0)+4(-1) = -12$$
$$2x = -8$$
$$x = -4$$

The solution set is $\{(-4,0,-1)\}$.

21. Let x represent the smallest number, y the second smallest number, and z the largest number.

$$\begin{pmatrix} x+y+z = 43 \\ x+y = z+3 \\ 2x+3y+4z = 141 \end{pmatrix}$$

- $x+y+z = 43$ ⟵ sum of three numbers is 43
- $x+y = z+3$ ⟵ sum of two smallest is three larger than the largest
- $2x+3y+4z = 141$ ⟵ twice the smallest plus three times the second number plus four times the largest is 141

Solving this system produces $x = 8$, $y = 15$, and $z = 20$.

25. Let x represent the measure of the smallest angle, y the measure of the second smallest angle, and z the measure of the largest angle.

$$\begin{pmatrix} x+y+z = 180 \\ z = 2x \\ x+z = 2y \end{pmatrix}$$

- $x+y+z = 180$ ⟵ sum of measures of angles of any triangle is 180°
- $z = 2x$ ⟵ measure of largest angle is twice the smallest
- $x+z = 2y$ ⟵ sum of smallest and largest is twice other angle

Solving this system produces $x = 40°$, $y = 60°$, and $z = 80°$.

29. Let x represent the number of Type A bird houses, y the number of Type B, and z the number of Type C.

$$\begin{pmatrix} .1x+.2y+.1z = 35 \\ .4x+.4y+.3z = 95 \\ .2x+.1y+.3z = 62.5 \end{pmatrix}$$

- $.1x+.2y+.1z = 35$ ⟵ cutting department spends 35 hours
- $.4x+.4y+.3z = 95$ ⟵ finishing department spends 95 hours
- $.2x+.1y+.3z = 62.5$ ⟵ assembling department spends 62.5 hours

Solving this system produces $x = 50$, $y = 75$, and $z = 150$.

Problem Set 7.3

13. augmented matrix ⟶ $\left[\begin{array}{cc|c} 3 & -4 & 33 \\ 1 & 7 & -39 \end{array}\right]$

Exchange the two rows.

$$\left[\begin{array}{cc|c} 1 & 7 & -39 \\ 3 & -4 & 33 \end{array}\right]$$

Multiply row 1 by -3 and add to row 2.

$$\left[\begin{array}{cc|c} 1 & 7 & -39 \\ 0 & -25 & 150 \end{array}\right]$$

Multiply row 2 by $-\frac{1}{25}$.

$$\left[\begin{array}{cc|c} 1 & 7 & -39 \\ 0 & 1 & -6 \end{array}\right]$$

Multiply row 2 by -7 and add to row 1.

$$\left[\begin{array}{cc|c} 1 & 0 & 3 \\ 0 & 1 & -6 \end{array}\right]$$

From this last matrix, we can see that the solution set is {(3,-6)}.

17. augmented matrix $\longrightarrow$ $\left[\begin{array}{cc|c} 3 & -5 & 39 \\ 2 & 7 & -67 \end{array}\right]$

Multiply row 1 by $\frac{1}{3}$.

$$\left[\begin{array}{cc|c} 1 & -\frac{5}{3} & 13 \\ 2 & 7 & -67 \end{array}\right]$$

Multiply row 1 by -2 and add to row 2.

$$\left[\begin{array}{cc|c} 1 & -\frac{5}{3} & 13 \\ 0 & \frac{31}{3} & -93 \end{array}\right]$$

Multiply row 2 by $\frac{3}{31}$.

$$\left[\begin{array}{cc|c} 1 & -\frac{5}{3} & 13 \\ 0 & 1 & -9 \end{array}\right]$$

Multiply row 2 by $\frac{5}{3}$ and add to row 1.

$$\left[\begin{array}{cc|c} 1 & 0 & -2 \\ 0 & 1 & -9 \end{array}\right]$$

The solution set is {(-2,-9)}.

21. augmented matrix $\longrightarrow$ $\left[\begin{array}{ccc|c} -2 & -5 & 3 & 11 \\ 1 & 3 & -3 & -12 \\ 3 & -2 & 5 & 31 \end{array}\right]$

Exchange rows 1 and 2.

$$\left[\begin{array}{ccc|c} 1 & 3 & -3 & -12 \\ -2 & -5 & 3 & 11 \\ 3 & -2 & 5 & 31 \end{array}\right]$$

Multiply row 1 by 2 and add to row 2. Also multiply row 1 by -3 and add to row 3.

$$\left[\begin{array}{ccc|c} 1 & 3 & -3 & -12 \\ 0 & 1 & -3 & -13 \\ 0 & -11 & 14 & 67 \end{array}\right]$$

Multiply row 2 by -3 and add to row 1. Also multiply row 2 by 11 and add to row 3.

$$\left[\begin{array}{ccc|c} 1 & 0 & 6 & 27 \\ 0 & 1 & -3 & -13 \\ 0 & 0 & -19 & -76 \end{array}\right]$$

From the last row we see that $-19z = -76$, which implies that $z = 4$. Then substitute 4 for z in the equation represented by the second row.

$$\begin{aligned} y-3z &= -13 \\ y-3(4) &= -13 \\ y &= -1 \end{aligned}$$

Finally, substitute 4 for z in the equation represented by the first row of the final matrix.

$$\begin{aligned} x+6z &= 27 \\ x+6(4) &= 27 \\ x &= 3 \end{aligned}$$

The solution set is $\{(3,-1,4)\}$.

25. augmented matrix $\longrightarrow$ $\left[\begin{array}{ccc|c} 1 & -1 & 2 & 1 \\ -3 & 4 & -1 & 4 \\ -1 & 2 & 3 & 6 \end{array}\right]$

Multiply row 1 by 3 and add to row 2. Also add row 1 to row 3.

$$\left[\begin{array}{ccc|c} 1 & -1 & 2 & 1 \\ 0 & 1 & 5 & 7 \\ 0 & 1 & 5 & 7 \end{array}\right]$$

Add row 2 to row 1. Also multiply row 2 by -1 and add to row 3.

$$\left[\begin{array}{ccc|c} 1 & 0 & 7 & 8 \\ 0 & 1 & 5 & 7 \\ 0 & 0 & 0 & 0 \end{array}\right]$$

The bottom row of zeros indicates a dependent system with infinitely many solutions. The second row represents the equation

$$y+5z = 7 \text{ or } y = 7-5z,$$

and the top row represents

$$x+7z = 8 \text{ or } x = 8-7z.$$

Therefore, if k represents any real number, the solution set can be represented by $\{(8-7k, 7-5k, k)\}$.

29.

augmented matrix $\longrightarrow$ $\left[\begin{array}{ccc|c} 2 & 3 & -1 & 7 \\ 3 & 4 & 5 & -2 \\ 5 & 1 & 3 & 13 \end{array}\right]$

Multiply row 1 by -1 and add to row 2.

$$\left[\begin{array}{ccc|c} 2 & 3 & -1 & 7 \\ 1 & 1 & 6 & -9 \\ 5 & 1 & 3 & 13 \end{array}\right]$$

Exchange rows 1 and 2.

$$\left[\begin{array}{ccc|c} 1 & 1 & 6 & -9 \\ 2 & 3 & -1 & 7 \\ 5 & 1 & 3 & 13 \end{array}\right]$$

Multiply row 1 by -2 and add to row 2. Also multiply row 1 by -5 and add to row 3.

$$\left[\begin{array}{ccc|c} 1 & 1 & 6 & -9 \\ 0 & 1 & -13 & 25 \\ 0 & -4 & -27 & 58 \end{array}\right]$$

Multiply row 2 by -1 and add to row 1. Also multiply row 2 by 4 and add to row 3.

$$\left[\begin{array}{ccc|c} 1 & 0 & 19 & -34 \\ 0 & 1 & -13 & 25 \\ 0 & 0 & -79 & 158 \end{array}\right]$$

From the last row we see that

$$-79z = 158$$

$$z = -2$$

Substituting -2 for z in the equation represented by the second row produces

$$y-13z = 25$$

$$y-13(-2) = 25$$

$$y = -1.$$

Substituting -2 for z in the equation represented by the first row produces

$$x+19z = -34$$
$$x+19(-2) = -34$$
$$x = 4.$$

The solution set is {(4,-1,-2)}.

33.

augmented matrix $\longrightarrow$ $\left[\begin{array}{cccc|c} 1 & 3 & -1 & 2 & -2 \\ 2 & 7 & 2 & -1 & 19 \\ -3 & -8 & 3 & 1 & -7 \\ 4 & 11 & -2 & -3 & 19 \end{array}\right]$

Multiply row 1 by -2 and add to row 2. Multiply row 1 by 3 and add to row 3. Multiply row 1 by -4 and add to row 4.

$$\left[\begin{array}{cccc|c} 1 & 3 & -1 & 2 & -2 \\ 0 & 1 & 4 & -5 & 23 \\ 0 & 1 & 0 & 7 & -13 \\ 0 & -1 & 2 & -11 & 27 \end{array}\right]$$

Multiply row 2 by -3 and add to row 1. Multiply row 2 by -1 and add to row 3. Add row 2 to row 4.

$$\left[\begin{array}{cccc|c} 1 & 0 & -13 & 17 & -71 \\ 0 & 1 & 4 & -5 & 23 \\ 0 & 0 & -4 & 12 & -36 \\ 0 & 0 & 6 & -16 & 50 \end{array}\right]$$

Multiply row 3 by $-\frac{1}{4}$.

$$\left[\begin{array}{cccc|c} 1 & 0 & -13 & 17 & -71 \\ 0 & 1 & 4 & -5 & 23 \\ 0 & 0 & 1 & -3 & 9 \\ 0 & 0 & 6 & -16 & 50 \end{array}\right]$$

Multiply row 3 by 13 and add to row 1. Multiply row 3 by -4 and add to row 2. Multiply row 3 by -6 and add to row 4.

$$\left[\begin{array}{cccc|c} 1 & 0 & 0 & -22 & 46 \\ 0 & 1 & 0 & 7 & -13 \\ 0 & 0 & 1 & -3 & 9 \\ 0 & 0 & 0 & 2 & -4 \end{array}\right]$$

From the last row we obtain

$$2x_4 = -4$$
$$x_4 = -2.$$

Substituting -2 for x_4 in the equations represented by rows 3, 2, and 1 produces the following results.

row 3	row 2	row 1
$x_3-3x_4 = 9$	$x_2+7x_4 = -13$	$x_1-22x_4 = 46$
$x_3-3(-2) = 9$	$x_2+7(-2) = -13$	$x_1-22(-2) = 46$
$x_3 = 3$	$x_2 = 1$	$x_1 = 2$

The solution set is $\{(2,1,3,-2)\}$.

37. The last row represents the statement $0(x_1)+0(x_2)+0(x_3)+0(x_4) = 1$, which is obviously false for all values of x_1, x_2, x_3, and x_4. Therefore, the system is inconsistent and the solution set is $\emptyset$.

41. From the matrix we see that $x_1+3x_2 = 9$ and $x_3 = 2$ and $x_4 = -3$. So let $x_2 = k$, where k is any real number. Then $x_1 +3x_2 = 9$ becomes $x_1+3k = 9$ or $x_1 = 9-3k$. So the solution set can be represented by

$\{(9-3k,k,2,-3)\}$.

45. augmented matrix $\longrightarrow$ $\left[\begin{array}{ccc|c} 2 & -4 & 3 & 8 \\ 3 & 5 & -1 & 7 \end{array}\right]$

Exchange rows.

$$\left[\begin{array}{ccc|c} 3 & 5 & -1 & 7 \\ 2 & -4 & 3 & 8 \end{array}\right]$$

Multiply row 2 by -1 and add to row 1.

$$\left[\begin{array}{ccc|c} 1 & 9 & -4 & -1 \\ 2 & -4 & 3 & 8 \end{array}\right]$$

Multiply row 1 by -2 and add to row 2.

$$\left[\begin{array}{ccc|c} 1 & 9 & -4 & -1 \\ 0 & -22 & 11 & 10 \end{array}\right]$$

Multiply row 2 by $-\frac{1}{22}$

$$\left[\begin{array}{ccc|c} 1 & 9 & -4 & -1 \\ 0 & 1 & -\frac{1}{2} & -\frac{5}{11} \end{array}\right]$$

Multiply row 2 by -9 and add to row 1.

$$\left[\begin{array}{ccc|c} 1 & 0 & \frac{1}{2} & \frac{34}{11} \\ 0 & 1 & -\frac{1}{2} & -\frac{5}{11} \end{array}\right]$$

The matrix represents the system

$$\begin{pmatrix} x + \frac{1}{2}z = \frac{34}{11} \\ y - \frac{1}{2}z = -\frac{5}{11} \end{pmatrix}.$$

The first equation can be written as

$$x = -\frac{1}{2}z + \frac{34}{11}$$

and the second as

$$y = \frac{1}{2}z - \frac{5}{11}.$$

Thus, if we let $z = k$, where k is any real number, the solution set can be represented by

$$\{(-\frac{1}{2}k + \frac{34}{11}, \frac{1}{2}k - \frac{5}{11}, k)\}.$$

Problem Set 7.4

1. $\begin{vmatrix} 4 & 3 \\ 2 & 7 \end{vmatrix} = 4(7)-3(2) = 28-6 = 22$

5. $\begin{vmatrix} 2 & -3 \\ 8 & -2 \end{vmatrix} = 2(-2)-(-3)(8) = -4+24 = 20$

9. $\begin{vmatrix} \frac{1}{2} & \frac{1}{3} \\ -3 & -6 \end{vmatrix} = \frac{1}{2}(-6)-(\frac{1}{3})(-3) = -3+1 = -2$

13. $\begin{vmatrix} 1 & 2 & -1 \\ 3 & 1 & 2 \\ 2 & 4 & 3 \end{vmatrix}$

To get some zeros in the first column, multiply row 1 by -3 and add to row 2, and multiply row 1 by -2 and add to row 3.

$$\begin{vmatrix} 1 & 2 & -1 \\ 0 & -5 & 5 \\ 0 & 0 & 5 \end{vmatrix}$$

Now expand about the first column.

$$1(-1)^{1+1}\begin{vmatrix} -5 & 5 \\ 0 & 5 \end{vmatrix} = 1(-25-0) = -25$$

17. $\begin{vmatrix} 6 & 12 & 3 \\ -1 & 5 & 1 \\ -3 & 6 & 2 \end{vmatrix}$

Multiply row 2 by 6 and add to row 1. Multiply row 2 by -1 and add to row 3.

$$\begin{vmatrix} 0 & 42 & 9 \\ -1 & 5 & 1 \\ 0 & -9 & -1 \end{vmatrix}$$

Now expand about the first column.

$$(-1)(-1)^{2+1}\begin{vmatrix} 42 & 9 \\ -9 & -1 \end{vmatrix} = (-1)(-1)(-42+81) = 39$$

21. $\begin{vmatrix} -3 & -2 & 1 \\ 5 & 0 & 6 \\ 2 & 1 & -4 \end{vmatrix}$

Multiply row 3 by 2 and add to row 1.

$$\begin{vmatrix} 1 & 0 & -7 \\ 5 & 0 & 6 \\ 2 & 1 & -4 \end{vmatrix}$$

Now expand about the second column.

$$1(-1)^{3+2}\begin{vmatrix} 1 & -7 \\ 5 & 6 \end{vmatrix} = (1)(-1)(6+35) = -41$$

25. Expand the given determinant about the third column.

$$(4)(-1)^{1+3}\begin{vmatrix} 40 & 2 \\ -16 & 6 \end{vmatrix} = (4)(1)(240+32) = 1088$$

29. $\begin{vmatrix} 1 & -2 & 3 & 2 \\ 2 & -1 & 0 & 4 \\ -3 & 4 & 0 & -2 \\ -1 & 1 & 1 & 5 \end{vmatrix}$

Multiply row 4 by -3 and add to row 1.

$$\begin{vmatrix} 4 & -5 & 0 & -13 \\ 2 & -1 & 0 & 4 \\ -3 & 4 & 0 & -2 \\ -1 & 1 & 1 & 5 \end{vmatrix}$$

Expand about the third column.

$$(1)(-1)^{4+3}\begin{vmatrix} 4 & -5 & -13 \\ 2 & -1 & 4 \\ -3 & 4 & -2 \end{vmatrix}$$

Now to expand the 3 X 3 determinant, let's get some zeros. Multiply row 2 by -5 and add to row 1. Also multiply row 2 by 4 and add to row 3.

$$\begin{vmatrix} -6 & 0 & -33 \\ 2 & -1 & 4 \\ 5 & 0 & 14 \end{vmatrix}$$

Now this determinant can be expanded about the second column.

$$(-1)(-1)^{2+2}\begin{vmatrix} -6 & -33 \\ 5 & 14 \end{vmatrix} = (-1)(1)(-84+165) = -81$$

Therefore, the original 4 X 4 determinant is

$$(1)(-1)^7(-81) = 81.$$

Problem Set 7.5

1. $D = \begin{vmatrix} 2 & -1 \\ 3 & 2 \end{vmatrix} = 4+3 = 7$

$D_x = \begin{vmatrix} -2 & -1 \\ 11 & 2 \end{vmatrix} = -4+11 = 7$

$D_y = \begin{vmatrix} 2 & -2 \\ 3 & 11 \end{vmatrix} = 22+6 = 28$

$x = \frac{D_x}{D} = \frac{7}{7} = 1$

$y = \frac{D_y}{D} = \frac{28}{7} = 4$

The solution set is $\{(1,4)\}$.

5. $D = \begin{vmatrix} 5 & -4 \\ -1 & 2 \end{vmatrix} = 10-4 = 6$

$D_x = \begin{vmatrix} 14 & -4 \\ -4 & 2 \end{vmatrix} = 28-16 = 12$

$D_y = \begin{vmatrix} 5 & 14 \\ -1 & -4 \end{vmatrix} = -20+14 = -6$

$x = \frac{D_x}{D} = \frac{12}{6} = 2$

$$y = \frac{D_y}{D} = \frac{-6}{6} = -1$$

The solution set is $\{(2,-1)\}$.

9. $$D = \begin{vmatrix} -4 & 3 \\ 4 & -6 \end{vmatrix} = 24-12 = 12$$

$$D_x = \begin{vmatrix} 3 & 3 \\ -5 & -6 \end{vmatrix} = -18+15 = -3$$

$$D_y = \begin{vmatrix} -4 & 3 \\ 4 & -5 \end{vmatrix} = 20-12 = 8$$

$$x = \frac{D_x}{D} = \frac{-3}{12} = -\frac{1}{4}$$

$$y = \frac{D_y}{D} = \frac{8}{12} = \frac{2}{3}$$

The solution set is $\{(-\frac{1}{4},\frac{2}{3})\}$.

13. $$D = \begin{vmatrix} -\frac{2}{3} & \frac{1}{2} \\ \frac{1}{3} & -\frac{3}{2} \end{vmatrix} = 1 - \frac{1}{6} = \frac{5}{6}$$

$$D_x = \begin{vmatrix} -7 & \frac{1}{2} \\ 6 & -\frac{3}{2} \end{vmatrix} = \frac{21}{2} - 3 = \frac{15}{2}$$

$$D_y = \begin{vmatrix} -\frac{2}{3} & -7 \\ \frac{1}{3} & 6 \end{vmatrix} = -4 + \frac{7}{3} = -\frac{5}{3}$$

$$x = \frac{D_x}{D} = \frac{\frac{15}{2}}{\frac{5}{6}} = 9$$

$$y = \frac{-\frac{5}{3}}{\frac{5}{6}} = -2$$

The solution set is $\{(9,-2)\}$.

17. $$D = \begin{vmatrix} 1 & -1 & 2 \\ 2 & 3 & -4 \\ -1 & 2 & -1 \end{vmatrix} = 13 \qquad D_x = \begin{vmatrix} -8 & -1 & 2 \\ 18 & 3 & -4 \\ 7 & 2 & -1 \end{vmatrix} = 0$$

$$D_y = \begin{vmatrix} 1 & -8 & 2 \\ 2 & 18 & -4 \\ -1 & 7 & -1 \end{vmatrix} = 26 \qquad D_z = \begin{vmatrix} 1 & -1 & -8 \\ 2 & 3 & 18 \\ -1 & 2 & 7 \end{vmatrix} = -39$$

$$x = \frac{D_x}{D} = \frac{0}{13} = 0; \quad y = \frac{D_y}{D} = \frac{26}{13} = 2; \quad z = \frac{D_z}{D} = \frac{-39}{13} = -3$$

The solution set is (0,2,-3) .

21.

$$D = \begin{vmatrix} 4 & 5 & -2 \\ 7 & -1 & 2 \\ 3 & 1 & 4 \end{vmatrix} = -154 \qquad D_x = \begin{vmatrix} -14 & 5 & -2 \\ 42 & -1 & 2 \\ 28 & 1 & 4 \end{vmatrix} = -616$$

$$D_y = \begin{vmatrix} 4 & -14 & -2 \\ 7 & 42 & 2 \\ 3 & 28 & 4 \end{vmatrix} = 616 \qquad D_z = \begin{vmatrix} 4 & 5 & -14 \\ 7 & -1 & 42 \\ 3 & 1 & 28 \end{vmatrix} = -770$$

$$x = \frac{D_x}{D} = \frac{-616}{-154} = 4; \quad y = \frac{D_y}{D} = \frac{616}{-154} = -4; \quad z = \frac{D_z}{D} = \frac{-770}{-154} = 5$$

The solution set is {(4,-4,5)}.

25.

$$D = \begin{vmatrix} 1 & 3 & -4 \\ 2 & -1 & 1 \\ 4 & 5 & -7 \end{vmatrix} = 0 \qquad D_x = \begin{vmatrix} -1 & 3 & -4 \\ 2 & -1 & 1 \\ 0 & 5 & -7 \end{vmatrix} = 0$$

$$D_y = \begin{vmatrix} 1 & -1 & -4 \\ 2 & 2 & 1 \\ 4 & 0 & -7 \end{vmatrix} = 0 \qquad D_z = \begin{vmatrix} 1 & 3 & -1 \\ 2 & -1 & 2 \\ 4 & 5 & 0 \end{vmatrix} = 0$$

The solution set has infinitely many solutions.

29.

$$D = \begin{vmatrix} 1 & -2 & 3 \\ -2 & 4 & -3 \\ 5 & -6 & 6 \end{vmatrix} = -12 \qquad D_x = \begin{vmatrix} 1 & -2 & 3 \\ -3 & 4 & -3 \\ 10 & -6 & 6 \end{vmatrix} = -36$$

$$D_y = \begin{vmatrix} 1 & 1 & 3 \\ -2 & -3 & -3 \\ 5 & 10 & 6 \end{vmatrix} = -6 \qquad D_z = \begin{vmatrix} 1 & -2 & 1 \\ -2 & 4 & -3 \\ 5 & -6 & 10 \end{vmatrix} = 4$$

$$x = \frac{D_x}{D} = \frac{-36}{-12} = 3; \quad y = \frac{D_y}{D} = \frac{-6}{-12} = \frac{1}{2}; \quad z = \frac{D_z}{D} = \frac{4}{-12} = -\frac{1}{3}$$

The solution set is $\{(3,\frac{1}{2}, -\frac{1}{3})\}$.

CHAPTER 8

Problem Set 8.1

1. $A+B = \begin{bmatrix} 1 & -2 \\ 3 & 4 \end{bmatrix} + \begin{bmatrix} 2 & -3 \\ 5 & -1 \end{bmatrix} = \begin{bmatrix} 3 & -5 \\ 8 & 3 \end{bmatrix}$

5. $4A-3B = 4\begin{bmatrix} 1 & -2 \\ 3 & 4 \end{bmatrix} - 3\begin{bmatrix} 2 & -3 \\ 5 & -1 \end{bmatrix}$

$= \begin{bmatrix} 4 & -8 \\ 12 & 16 \end{bmatrix} - \begin{bmatrix} 6 & -9 \\ 15 & -3 \end{bmatrix}$

$= \begin{bmatrix} -2 & 1 \\ -3 & 19 \end{bmatrix}$

9. $2D-4E = 2\begin{bmatrix} -2 & 3 \\ 5 & -4 \end{bmatrix} - 4\begin{bmatrix} 2 & 5 \\ 7 & 3 \end{bmatrix}$

$= \begin{bmatrix} -4 & 6 \\ 10 & -8 \end{bmatrix} - \begin{bmatrix} 8 & 20 \\ 28 & 12 \end{bmatrix} = \begin{bmatrix} -12 & -14 \\ -18 & -20 \end{bmatrix}$

13. $AB = \begin{bmatrix} 1 & -1 \\ 2 & -2 \end{bmatrix}\begin{bmatrix} 3 & -4 \\ -1 & 2 \end{bmatrix} = \begin{bmatrix} 1(3)+(-1)(-1) & 1(-4)+(-1)(2) \\ 2(3)+(-2)(-1) & 2(-4)+(-2)(2) \end{bmatrix}$

$= \begin{bmatrix} 4 & -6 \\ 8 & -12 \end{bmatrix}$

$BA = \begin{bmatrix} 3 & -4 \\ -1 & 2 \end{bmatrix}\begin{bmatrix} 1 & -1 \\ 2 & -2 \end{bmatrix} = \begin{bmatrix} 3(1)+(-4)(2) & 3(-1)+(-4)(-2) \\ (-1)(1)+(2)(2) & (-1)(-1)+(2)(-2) \end{bmatrix}$

$= \begin{bmatrix} -5 & 5 \\ 3 & -3 \end{bmatrix}$

17. $AB = \begin{bmatrix} 2 & -4 \\ 1 & -2 \end{bmatrix}\begin{bmatrix} 1 & -2 \\ -3 & 6 \end{bmatrix} = \begin{bmatrix} 2(1)+(-4)(-3) & 2(-2)+(-4)(6) \\ 1(1)+(-2)(-3) & 1(-2)+(-2)(6) \end{bmatrix}$

$= \begin{bmatrix} 14 & -28 \\ 7 & -14 \end{bmatrix}$

$BA = \begin{bmatrix} 1 & -2 \\ -3 & 6 \end{bmatrix}\begin{bmatrix} 2 & -4 \\ 1 & -2 \end{bmatrix} = \begin{bmatrix} 1(2)+(-2)(1) & 1(-4)+(-2)(-2) \\ -3(2)+6(1) & -3(-4)+6(-2) \end{bmatrix}$

$= \begin{bmatrix} 0 & 0 \\ 0 & 0 \end{bmatrix}$

21. $AB = \begin{bmatrix} 2 & -1 \\ -5 & 3 \end{bmatrix}\begin{bmatrix} 3 & 1 \\ 5 & 2 \end{bmatrix} = \begin{bmatrix} 2(3)+(-1)(5) & 2(-1)+(-1)(2) \\ -5(3)+3(5) & -5(1)+3(2) \end{bmatrix}$

$= \begin{bmatrix} 1 & 0 \\ 0 & 1 \end{bmatrix}$

$$BA = \begin{bmatrix} 3 & 1 \\ 5 & 2 \end{bmatrix}\begin{bmatrix} 2 & -1 \\ -5 & 3 \end{bmatrix} = \begin{bmatrix} 3(2)+1(-5) & 3(-1)+1(3) \\ 5(2)+2(-5) & 5(-1)+2(3) \end{bmatrix}$$

$$= \begin{bmatrix} 1 & 0 \\ 0 & 1 \end{bmatrix}$$

25. $$AB = \begin{bmatrix} 5 & 6 \\ 2 & 3 \end{bmatrix}\begin{bmatrix} 1 & -2 \\ -\frac{2}{3} & \frac{5}{3} \end{bmatrix} = \begin{bmatrix} 5(1)+6(-\frac{2}{3}) & 5(-2)+6(\frac{5}{3}) \\ 2(1)+3(-\frac{2}{3}) & 2(-2)+3(\frac{5}{3}) \end{bmatrix}$$

$$= \begin{bmatrix} 1 & 0 \\ 0 & 1 \end{bmatrix}$$

$$BA = \begin{bmatrix} 1 & -2 \\ -\frac{2}{3} & \frac{5}{3} \end{bmatrix}\begin{bmatrix} 5 & 6 \\ 2 & 3 \end{bmatrix} = \begin{bmatrix} 1(5)+(-2)(2) & 1(6)+(-2)(3) \\ -\frac{2}{3}(5)+\frac{5}{3}(2) & -\frac{2}{3}(6)+\frac{5}{3}(3) \end{bmatrix}$$

$$= \begin{bmatrix} 1 & 0 \\ 0 & 1 \end{bmatrix}$$

29. $$AD = \begin{bmatrix} -2 & 3 \\ 5 & 4 \end{bmatrix}\begin{bmatrix} 1 & 1 \\ 1 & 1 \end{bmatrix} = \begin{bmatrix} -2(1)+3(1) & -2(1)+3(1) \\ 5(1)+4(1) & 5(1)+4(1) \end{bmatrix}$$

$$= \begin{bmatrix} 1 & 1 \\ 9 & 9 \end{bmatrix}$$

$$DA = \begin{bmatrix} 1 & 1 \\ 1 & 1 \end{bmatrix}\begin{bmatrix} -2 & 3 \\ 5 & 4 \end{bmatrix} = \begin{bmatrix} 1(-2)+1(5) & 1(3)+1(4) \\ 1(-2)+1(5) & 1(3)+1(4) \end{bmatrix}$$

$$= \begin{bmatrix} 3 & 7 \\ 3 & 7 \end{bmatrix}$$

33. $$(A+B)C = \left(\begin{bmatrix} 2 & 4 \\ 5 & -3 \end{bmatrix} + \begin{bmatrix} -2 & 3 \\ -1 & 2 \end{bmatrix}\right)\left(\begin{bmatrix} 2 & 1 \\ 3 & 7 \end{bmatrix}\right)$$

$$= \begin{bmatrix} 0 & 7 \\ 4 & -1 \end{bmatrix}\begin{bmatrix} 2 & 1 \\ 3 & 7 \end{bmatrix} = \begin{bmatrix} 21 & 49 \\ 5 & -3 \end{bmatrix}$$

$$AC + BC = \begin{bmatrix} 2 & 4 \\ 5 & -3 \end{bmatrix}\begin{bmatrix} 2 & 1 \\ 3 & 7 \end{bmatrix} + \begin{bmatrix} -2 & 3 \\ -1 & 2 \end{bmatrix}\begin{bmatrix} 2 & 1 \\ 3 & 7 \end{bmatrix}$$

$$= \begin{bmatrix} 16 & 30 \\ 1 & -16 \end{bmatrix} + \begin{bmatrix} 5 & 19 \\ 4 & 13 \end{bmatrix} = \begin{bmatrix} 21 & 49 \\ 5 & -3 \end{bmatrix}$$

37. $$A+(-A) = \begin{bmatrix} a_{11} & a_{12} \\ a_{21} & a_{22} \end{bmatrix} + \begin{bmatrix} -a_{11} & -a_{12} \\ -a_{21} & -a_{22} \end{bmatrix} = \begin{bmatrix} 0 & 0 \\ 0 & 0 \end{bmatrix}$$

45. $A^2 = A \cdot A = \begin{bmatrix} 1 & -1 \\ 2 & 3 \end{bmatrix}\begin{bmatrix} 1 & -1 \\ 2 & 3 \end{bmatrix} = \begin{bmatrix} -1 & -4 \\ 8 & 7 \end{bmatrix}$

$A^3 = A^2 \cdot A = \begin{bmatrix} -1 & -4 \\ 8 & 7 \end{bmatrix}\begin{bmatrix} 1 & -1 \\ 2 & 3 \end{bmatrix} = \begin{bmatrix} -9 & -11 \\ 22 & 13 \end{bmatrix}$

Problem Set 8.2

1. $|A| = 15-14 = 1$ $\quad A^{-1} = \frac{1}{1}\begin{bmatrix} 3 & -7 \\ -2 & 5 \end{bmatrix} = \begin{bmatrix} 3 & -7 \\ -2 & 5 \end{bmatrix}$

5. $|A| = -4-6 = -10$ $\quad A^{-1} = -\frac{1}{10}\begin{bmatrix} 4 & -2 \\ -3 & -1 \end{bmatrix} = \begin{bmatrix} -\frac{2}{5} & \frac{1}{5} \\ \frac{3}{10} & \frac{1}{10} \end{bmatrix}$

9. $|A| = -15-(-8) = -7$ $\quad A^{-1} = -\frac{1}{7}\begin{bmatrix} 5 & -2 \\ 4 & -3 \end{bmatrix} = \begin{bmatrix} -\frac{5}{7} & \frac{2}{7} \\ -\frac{4}{7} & \frac{3}{7} \end{bmatrix}$

13. $|A| = 8-3 = 5$ $\quad A^{-1} = \frac{1}{5}\begin{bmatrix} -4 & 3 \\ 1 & -2 \end{bmatrix} = \begin{bmatrix} -\frac{4}{5} & \frac{3}{5} \\ \frac{1}{5} & -\frac{2}{5} \end{bmatrix}$

17. $|A| = -1-1 = -2$ $\quad A^{-1} = -\frac{1}{2}\begin{bmatrix} -1 & -1 \\ -1 & 1 \end{bmatrix} = \begin{bmatrix} \frac{1}{2} & \frac{1}{2} \\ \frac{1}{2} & -\frac{1}{2} \end{bmatrix}$

21. $AB = \begin{bmatrix} -3 & -4 \\ 2 & 1 \end{bmatrix}\begin{bmatrix} 4 \\ -3 \end{bmatrix} = \begin{bmatrix} -3(4)+(-4)(-3) \\ 2(4)+1(-3) \end{bmatrix} = \begin{bmatrix} 0 \\ 5 \end{bmatrix}$

25. $AB = \begin{bmatrix} -2 & -3 \\ -5 & -6 \end{bmatrix}\begin{bmatrix} 5 \\ -2 \end{bmatrix} = \begin{bmatrix} -2(5)+(-3)(-2) \\ -5(5)+(-6)(-2) \end{bmatrix} = \begin{bmatrix} -4 \\ -13 \end{bmatrix}$

Examples 29, 33, and 37 are solved using the technique demonstrated in Examples 4 and 5 of Section 8.2 in the text.

29. By letting $A = \begin{bmatrix} 4 & -3 \\ -3 & 2 \end{bmatrix}$, $X = \begin{bmatrix} x \\ y \end{bmatrix}$, and $B = \begin{bmatrix} -23 \\ 16 \end{bmatrix}$, the given system is represented by the matrix equation $AX = B$. Therefore, $X = A^{-1}B$.

$|A| = 8-9 = -1$

$A^{-1} = \frac{1}{-1}\begin{bmatrix} 2 & 3 \\ 3 & 4 \end{bmatrix} = \begin{bmatrix} -2 & -3 \\ -3 & -4 \end{bmatrix}$

$X = \begin{bmatrix} -2 & -3 \\ -3 & -4 \end{bmatrix}\begin{bmatrix} -23 \\ 16 \end{bmatrix} = \begin{bmatrix} -2(-23)+(-3)(16) \\ -3(-23)+(-4)(16) \end{bmatrix} = \begin{bmatrix} -2 \\ 5 \end{bmatrix}$

The solution set is $\{(-2,5)\}$.

33. $A = \begin{bmatrix} 3 & -5 \\ 4 & -3 \end{bmatrix}$; $|A| = -9-(-20) = 11$

$$A^{-1} = \frac{1}{11}\begin{bmatrix} -3 & 5 \\ -4 & 3 \end{bmatrix} = \begin{vmatrix} -\frac{3}{11} & \frac{5}{11} \\ -\frac{4}{11} & \frac{3}{11} \end{vmatrix}$$

$$X = \begin{bmatrix} -\frac{3}{11} & \frac{5}{11} \\ -\frac{4}{11} & \frac{3}{11} \end{bmatrix}\begin{bmatrix} 2 \\ -1 \end{bmatrix} = \begin{bmatrix} -\frac{3}{11}(2) + \frac{5}{11}(-1) \\ -\frac{4}{11}(2) + \frac{3}{11}(-1) \end{bmatrix} = \begin{bmatrix} -1 \\ -1 \end{bmatrix}$$

The solution set is $\{(-1,-1)\}$.

37. $A = \begin{bmatrix} 3 & 2 \\ 30 & -18 \end{bmatrix}$; $|A| = -54-60 = -114$

$$A^{-1} = \frac{1}{-114}\begin{bmatrix} -18 & -2 \\ -30 & 3 \end{bmatrix} = \begin{bmatrix} \frac{3}{19} & \frac{1}{57} \\ \frac{5}{19} & -\frac{1}{38} \end{bmatrix}$$

$$X = \begin{bmatrix} \frac{3}{19} & \frac{1}{57} \\ \frac{5}{19} & -\frac{1}{38} \end{bmatrix}\begin{bmatrix} 0 \\ -19 \end{bmatrix} = \begin{bmatrix} \frac{3}{19}(0)+ \frac{1}{57}(-19) \\ \frac{5}{19}(0) +(-\frac{1}{38})(-19) \end{bmatrix} = \begin{bmatrix} -\frac{1}{3} \\ \frac{1}{2} \end{bmatrix}$$

The solution set is $\{(-\frac{1}{3},\frac{1}{2})\}$.

Problem Set 8.3

1 and 5. To find A + B, add the corresponding elements of A and B.

To find A - B, treat A - B as A + (-B). Thus, we add the elements of A and the corresponding elements of -B.

To find 2A + 3B, double each element of A, triple each element of B, and then add corresponding elements.

To find 4A - 2B, treat 4A - 2B as 4A + (-2B).

9.

$$AB = \begin{bmatrix} 2 & -1 \\ 0 & -4 \\ -5 & 3 \end{bmatrix}\begin{bmatrix} 5 & -2 & 6 \\ -1 & 4 & -2 \end{bmatrix}$$

$$= \begin{bmatrix} 2(5)+(-1)(-1) & 2(-2)+(-1)(4) & 2(6)+(-1)(-2) \\ 0(5)+(-4)(-1) & 0(-2)+(-4)(4) & 0(6)+(-4)(-2) \\ -5(5)+3(-1) & -5(-2)+3(4) & -5(6)+3(-2) \end{bmatrix}$$

$$= \begin{bmatrix} 11 & -8 & 14 \\ 4 & -16 & 8 \\ -28 & 22 & -36 \end{bmatrix}$$

$$BA = \begin{bmatrix} 5 & -2 & 6 \\ -1 & 4 & -2 \end{bmatrix} \begin{bmatrix} 2 & -1 \\ 0 & -4 \\ -5 & 3 \end{bmatrix}$$

$$= \begin{bmatrix} 5(2)+(-2)(0)+6(-5) & 5(-1)+(-2)(-4)+6(3) \\ -1(2)+4(0)+(-2)(-5) & -1(-1)+4(-4)+(-2)(3) \end{bmatrix} = \begin{bmatrix} -20 & 21 \\ 8 & -21 \end{bmatrix}$$

13. $$AB = \begin{bmatrix} 1 & -1 & 2 \\ 0 & 1 & -2 \\ 3 & 1 & 4 \end{bmatrix} \begin{bmatrix} 2 & 3 & -1 \\ 4 & 0 & 2 \\ -5 & 1 & -1 \end{bmatrix}$$

$$= \begin{bmatrix} 1(2)+(-1)(4)+2(-5) & 1(3)+(-1)(0)+2(1) & 1(-1)+(-1)(2)+2(-1) \\ 0(2)+1(4)+(-2)(-5) & 0(3)+1(0)+(-2)(1) & 0(-1)+1(2)+(-2)(-1) \\ 3(2)+1(4)+4(-5) & 3(3)+1(0)+4(1) & 3(-1)+1(2)+4(-1) \end{bmatrix}$$

$$= \begin{bmatrix} -12 & 5 & -5 \\ 14 & -2 & 4 \\ -10 & 13 & -5 \end{bmatrix}$$

$$BA = \begin{bmatrix} 2 & 3 & -1 \\ 4 & 0 & 2 \\ -5 & 1 & -1 \end{bmatrix} \begin{bmatrix} 1 & -1 & 2 \\ 0 & 1 & -2 \\ 3 & 1 & 4 \end{bmatrix}$$

$$= \begin{bmatrix} 2(1)+3(0)+(-1)(3) & 2(-1)+3(1)+(-1)(1) & 2(2)+3(-2)+(-1)(4) \\ 4(1)+0(0)+2(3) & 4(-1)+0(1)+2(1) & 4(2)+0(-2)+2(4) \\ -5(1)+1(0)+(-1)(3) & -5(-1)+1(1)+(-1)(1) & -5(2)+1(-2)+(-1)(4) \end{bmatrix}$$

$$= \begin{bmatrix} -1 & 0 & -6 \\ 10 & -2 & 16 \\ -8 & 5 & -16 \end{bmatrix}$$

17. AB does not exist since the number of columns of A does not equal the number of rows of B.

$$BA = \begin{bmatrix} 3 & -2 \\ 1 & 0 \\ -1 & 4 \end{bmatrix} \begin{bmatrix} 2 \\ -7 \end{bmatrix} = \begin{bmatrix} 3(2)+(-2)(-7) \\ 1(2)+0(-7) \\ -1(2)+4(-7) \end{bmatrix} = \begin{bmatrix} 20 \\ 2 \\ -30 \end{bmatrix}$$

21. $$\left[\begin{array}{cc|cc} 1 & 3 & 1 & 0 \\ 4 & 2 & 0 & 1 \end{array}\right]$$

Multiply row 1 by -4 and add to row 2.

$$\left[\begin{array}{cc|cc} 1 & 3 & 1 & 0 \\ 0 & -10 & -4 & 1 \end{array}\right]$$

Multiply row 2 by $-\frac{1}{10}$.

$$\left[\begin{array}{cc|cc} 1 & 3 & 1 & 0 \\ 0 & 1 & \frac{2}{5} & -\frac{1}{10} \end{array}\right]$$

Multiply row 2 by -3 and add to row 1.

$$\left[\begin{array}{cc|cc} 1 & 0 & -\frac{1}{5} & \frac{3}{10} \\ 0 & 1 & \frac{2}{5} & -\frac{1}{10} \end{array}\right]$$

The multiplicative inverse is $\begin{bmatrix} -\frac{1}{5} & \frac{3}{10} \\ \frac{2}{5} & -\frac{1}{10} \end{bmatrix}$.

25. $$\left[\begin{array}{cc|cc} -2 & 1 & 1 & 0 \\ 3 & -4 & 0 & 1 \end{array}\right]$$

Add row 2 to row 1.

$$\left[\begin{array}{cc|cc} 1 & -3 & 1 & 1 \\ 3 & -4 & 0 & 1 \end{array}\right]$$

Multiply row 1 by -3 and add to row 2.

$$\left[\begin{array}{cc|cc} 1 & -3 & 1 & 1 \\ 0 & 5 & -3 & -2 \end{array}\right]$$

Multiply row 2 by $\frac{1}{5}$.

$$\left[\begin{array}{cc|cc} 1 & -3 & 1 & 1 \\ 0 & 1 & -\frac{3}{5} & -\frac{2}{5} \end{array}\right]$$

Multiply row 2 by 3 and add to row 1.

$$\left[\begin{array}{cc|cc} 1 & 0 & -\frac{4}{5} & -\frac{1}{5} \\ 0 & 1 & -\frac{3}{5} & -\frac{2}{5} \end{array}\right]$$

The multiplicative inverse is $\begin{bmatrix} -\frac{4}{5} & -\frac{1}{5} \\ -\frac{3}{5} & -\frac{2}{5} \end{bmatrix}$.

29. $$\left[\begin{array}{ccc|ccc} 1 & -2 & 1 & 1 & 0 & 0 \\ -2 & 5 & 3 & 0 & 1 & 0 \\ 3 & -5 & 7 & 0 & 0 & 1 \end{array}\right]$$

Multiply row 1 by 2 and add to row 2. Multiply row 1 by -3 and add to row 3.

$$\left[\begin{array}{ccc|ccc} 1 & -2 & 1 & 1 & 0 & 0 \\ 0 & 1 & 5 & 2 & 1 & 0 \\ 0 & 1 & 4 & -3 & 0 & 1 \end{array}\right]$$

Multiply row 2 by 2 and add to row 1. Multiply row 2 by -1 and add to row 3.

$$\left[\begin{array}{ccc|ccc} 1 & 0 & 11 & 5 & 2 & 0 \\ 0 & 1 & 5 & 2 & 1 & 0 \\ 0 & 0 & -1 & -5 & -1 & 1 \end{array}\right]$$

Multiply row 3 by -1

$$\left[\begin{array}{ccc|ccc} 1 & 0 & 11 & 5 & 2 & 0 \\ 0 & 1 & 5 & 2 & 1 & 0 \\ 0 & 0 & 1 & 5 & 1 & -1 \end{array}\right]$$

Multiply row 3 by -11 and add to row 1. Multiply row 3 by -5 and add to row 2.

$$\left[\begin{array}{ccc|ccc} 1 & 0 & 0 & -50 & -9 & 11 \\ 0 & 1 & 0 & -23 & -4 & 5 \\ 0 & 0 & 1 & 5 & 1 & -1 \end{array}\right]$$

The multiplicative inverse is $\begin{bmatrix} -50 & -9 & 11 \\ -23 & -4 & 5 \\ 5 & 1 & -1 \end{bmatrix}$.

33. $\left[\begin{array}{ccc|ccc} 1 & 2 & 3 & 1 & 0 & 0 \\ -3 & -4 & 3 & 0 & 1 & 0 \\ 2 & 4 & -1 & 0 & 0 & 1 \end{array}\right]$

Multiply row 1 by 3 and add to row 2. Multiply row 1 by -2 and add to row 3.

$$\left[\begin{array}{ccc|ccc} 1 & 2 & 3 & 1 & 0 & 0 \\ 0 & 2 & 12 & 3 & 1 & 0 \\ 0 & 0 & -7 & -2 & 0 & 1 \end{array}\right]$$

Multiply row 2 by $\frac{1}{2}$.

$$\left[\begin{array}{ccc|ccc} 1 & 2 & 3 & 1 & 0 & 0 \\ 0 & 1 & 6 & \frac{3}{2} & \frac{1}{2} & 0 \\ 0 & 0 & -7 & -2 & 0 & 1 \end{array}\right]$$

Multiply row 2 by -2 and add to row 1.

$$\left[\begin{array}{ccc|ccc} 1 & 0 & -9 & -2 & -1 & 0 \\ 0 & 1 & 6 & \frac{3}{2} & \frac{1}{2} & 0 \\ 0 & 0 & -7 & -2 & 0 & 1 \end{array}\right]$$

Multiply row 3 by $-\frac{1}{7}$.

$$\left[\begin{array}{ccc|ccc} 1 & 0 & -9 & -2 & -1 & 0 \\ 0 & 1 & 6 & \frac{3}{2} & \frac{1}{2} & 0 \\ 0 & 0 & 1 & \frac{2}{7} & 0 & -\frac{1}{7} \end{array}\right]$$

Multiply row 3 by 9 and add to row 1. Multiply row 3 by -6 and add to row 2.

$$\left[\begin{array}{ccc|ccc} 1 & 0 & 0 & \frac{4}{7} & -1 & -\frac{9}{7} \\ 0 & 1 & 0 & -\frac{3}{14} & \frac{1}{2} & \frac{6}{7} \\ 0 & 0 & 1 & \frac{2}{7} & 0 & -\frac{1}{7} \end{array}\right]$$

The multiplicative inverse is $\begin{bmatrix} \frac{4}{7} & -1 & -\frac{9}{7} \\ -\frac{3}{14} & \frac{1}{2} & \frac{6}{7} \\ \frac{2}{7} & 0 & -\frac{1}{7} \end{bmatrix}$

37. The multiplicative inverse of the coefficient matrix was found in Problem 23. Therefore, the solution of the system can be found as follows.

$$X = \begin{bmatrix} 4 & -1 \\ -7 & 2 \end{bmatrix} \begin{bmatrix} -4 \\ -13 \end{bmatrix} = \begin{bmatrix} -3 \\ 2 \end{bmatrix}$$

↑ inverse of coefficient matrix

The solution set is $\{(-3,2)\}$.

41. The multiplicative inverse of the coefficient matrix was found in Problem 27.

$$X = \begin{bmatrix} \frac{7}{2} & -3 & \frac{1}{2} \\ -\frac{1}{2} & 0 & \frac{1}{2} \\ -\frac{1}{2} & 1 & -\frac{1}{2} \end{bmatrix} \begin{bmatrix} -2 \\ -3 \\ -6 \end{bmatrix} = \begin{bmatrix} -1 \\ -2 \\ 1 \end{bmatrix}$$

↑ inverse of coefficient matrix

The solution set is $\{(-1,-2,1)\}$.

45. The multiplicative inverse of the coefficient matrix was found in Problem 33.

$$X = \underbrace{\begin{bmatrix} \frac{4}{7} & -1 & -\frac{9}{7} \\ -\frac{3}{14} & \frac{1}{2} & \frac{6}{7} \\ \frac{2}{7} & 0 & -\frac{1}{7} \end{bmatrix}}_{\text{inverse of coefficient matrix}} \begin{bmatrix} 2 \\ 0 \\ 4 \end{bmatrix} = \begin{bmatrix} -4 \\ 3 \\ 0 \end{bmatrix}$$

The solution set is $\{(-4,3,0)\}$.

49. (a) $\begin{bmatrix} -1 & 0 \\ 0 & 1 \end{bmatrix} \begin{bmatrix} x \\ y \end{bmatrix} = \begin{bmatrix} -x \\ y \end{bmatrix}$

y

(-x,y) (x,y)

x

The points (x,y) and (-x,y) are y-axis reflections of each other.

(c) $\begin{bmatrix} 0 & -1 \\ 1 & 0 \end{bmatrix} \begin{bmatrix} x \\ y \end{bmatrix} = \begin{bmatrix} -y \\ x \end{bmatrix}$

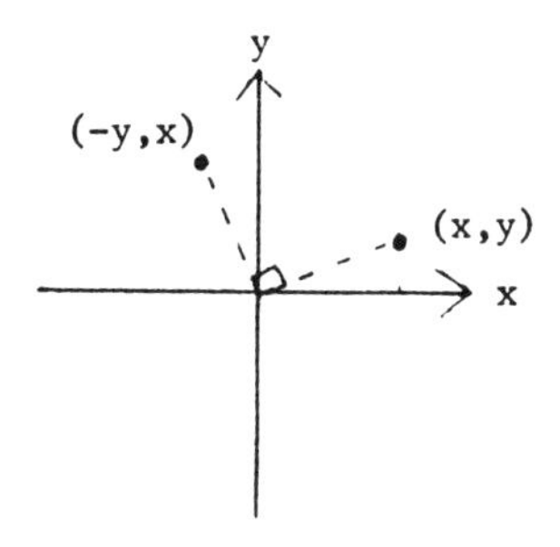

The point (-y,x) is a 90° counterclockwise rotation of (x,y).

Problem Set 8.4

1. The graph of $x+y > 3$ consists of all points above the line $x+y = 3$. The graph of $x-y > 1$ consists of all points below the line $x-y = 1$. Therefore, the solution set of the system is the intersection of those two sets of points.

5. The graph of $2x+3y \leq 6$ consists of all points on or below the line $2x+3y = 6$. The graph of $3x-2y \leq 6$ consists of all points on or above the line $3x-2y = 6$. Therefore, the solution set of the system is the intersection of those two sets of points.

9. The graph of $x+2y > -2$ consists of all points above the line $x+2y = -2$. The graph of $x-y < -3$ consists of all points above the line $x-y = -3$. Thus, the solution set of the system is the intersection of those two sets of points.

13. The graph of $x-y > 2$ consists of all points below the line $x-y = 2$. The graph of $x-y > -1$ consists of all points below the line $x-y = -1$. Thus, the solution set of the system is the intersection of those two sets of points.

17. The graph of $y < x$ consists of all points below the line $y = x$. The graph of $y > x+3$ consists of all points above the line $y = x+3$. Note that the two lines are parallel. The solution set of the system is the intersection of the two sets of points, which is the empty set.

21. The graph of $x+y \leq 4$ consists of all points below the line $x+y = 4$. The graph of $2x+y \leq 6$ consists of all points below the line $2x+y = 6$. The intersection of these two sets of points along with $x \geq 0$ and $y \geq 0$ produces the solution set of the given system.

25. Remember that the maximum and minimum values occur at vertices of the region. Therefore, we simply need to compute the functional values at the vertices and then observe the largest and smallest values.

$$f(x,y) = 3x+5y$$

$f(1,1) = 3(1)+5(1) = 8 \leftarrow$ minimum

$f(2,4) = 3(2)+5(4) = 26$

$f(4,8) = 3(4)+5(8) = 52 \leftarrow$ maximum

$f(5,2) = 3(5)+5(2) = 25$

29. First, let's sketch the lines and determine the region under consideration.

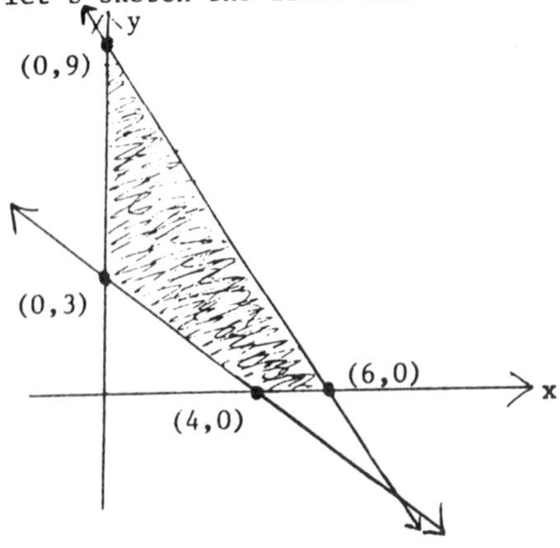

Now we can determine the functional values of $f(x,y) = 3x+7y$ at the vertices.

$f(0,3) = 3(0)+7(3) = 21$

$f(0,9) = 3(0)+7(9) = 63$ ⟵ maximum

$f(6,0) = 3(6)+7(0) = 18$

$f(4,0) = 3(4)+7(0) = 12$

33. First, let's sketch the lines and determine the region under consideration.

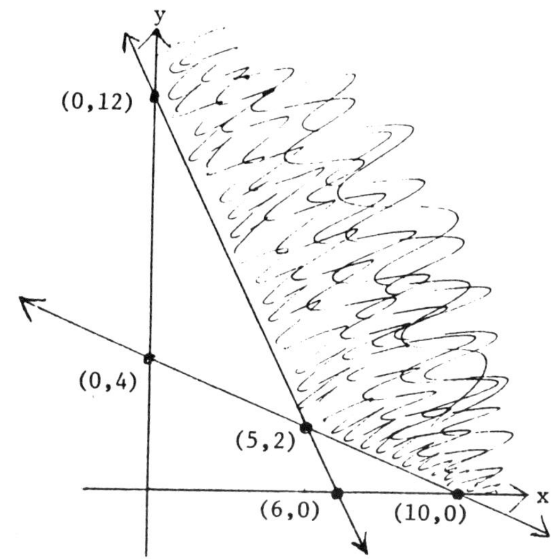

The point (5,2) is found by solving the system $\begin{pmatrix} 2x+y = 12 \\ 2x+5y = 20 \end{pmatrix}$.

Now we can determine the functional value of $f(x,y) = .2x+.5y$ at the vertices (0,12), (5,2), and (10,0).

$f(0,12) = .2(0)+.5(12) = 6$

$f(5,2) = .2(5)+.5(2) = 2$ ⟵ minimum

$f(10,0) = .2(10)+.5(0) = 2$ ⟵ minimum

37. Let x represent the amount invested in the conservative stock and y the amount invested in the speculative stock.

$x+y \leq 10,000$ ⟵ She wants to invest up to \$10,000.

$x \geq 2,000$ ⟵ at least \$2000 in conservative stock

$y \leq 6,000$ ⟵ no more than \$6000 in speculative stock

$y \leq x$ ⟵ speculative not to exceed conservative investment

Now let's determine the set of feasible solutions.

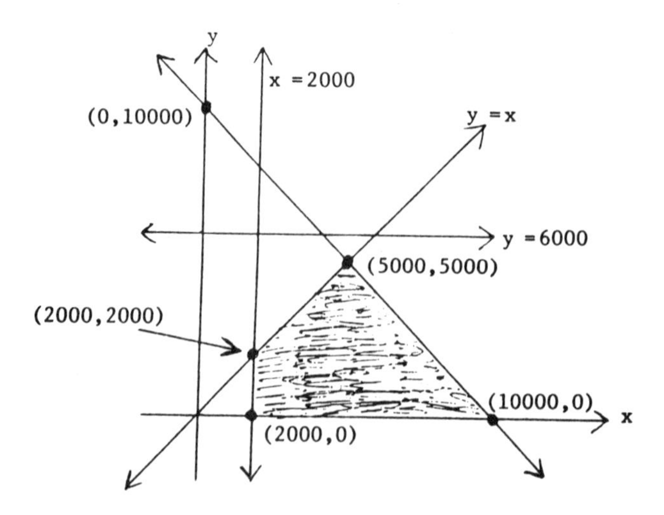

Since the conservative pays 9% return and the speculative a 12% return, the function to be maximized is $f(x,y) = .09x + .12y$. Therefore, evaluating this function at the vertices of the shaded region produces the following results.

$$f(2000,0) = .09(2000) + .12(0) = 180$$

$$f(2000,2000) = .09(2000) + .12(2000) = 420$$

$$f(5000,5000) = .09(5000) + .12(5000) = 1050$$

$$f(10000,0) = .09(10000) + .12(0) = 900$$

Therefore, \$5000 should be invested at 9% and \$5000 at 12%.

41. Let x be the number of units of product A and y the number of units of product B.

$x+y \le 40$ ⟵ Machine I is available for no more than 40 hours.

$2x+y \le 40$ ⟵ Machine II is available for no more than 40 hours.

$x+3y \le 60$ ⟵ Machine III is available for no more than 60 hours.

$x \ge 0$

$y \ge 0$

Now let's determine the set of feasible solutions.

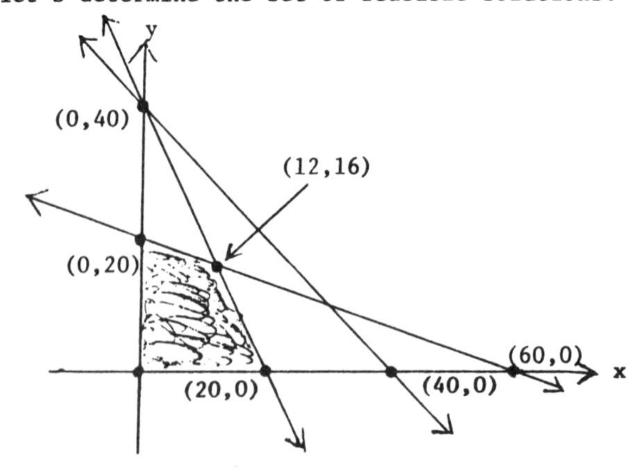

The profit function to be maximized is $f(x,y) = 2.75x + 3.50y$.

$$f(0,0) = 2.75(0) + 3.50(0) = 0$$

$$f(0,20) = 2.75(0) + 3.50(20) = 70$$

$$f(12,16) = 2.75(12) + 3.50(16) = 89$$

$$f(20,0) = 2.75(20) + 3.50(0) = 55$$

She produced 12 units of product A and 16 units of product B.

CHAPTER 9

Problem Set 9.1

1. The equation $y^2 = 8x$ is of the form $y^2 = 4px$; thus, $4p = 8$ and $p = 2$. So the focus is at (2,0) and the equation of the directrix is $x = -2$. Since $|4p| = |8| = 8$, the ends of the latus rectum are at (2,4) and (2,-4). The parabola can be sketched using the points (0,0),(2,4), and (2,-4).

5. The equation $y^2 = -2x$ is of the form $y^2 = 4px$; thus, $4p = -2$ and $p = -\frac{1}{2}$. So the focus is at $(-\frac{1}{2},0)$ and the equation of the directrix is $x = \frac{1}{2}$. Since $|4p| = |-2| = 2$, the ends of the latus rectum are at $(-\frac{1}{2},1)$ and $(-\frac{1}{2},-1)$. The parabola can be sketched using the points (0,0),$(-\frac{1}{2},1)$, and $(-\frac{1}{2},-1)$.

9. $$\begin{aligned} x^2-4y+8 &= 0 \\ x^2 &= 4y-8 \\ x^2 &= 4(y-2) \end{aligned}$$

   Now let's compare this equation to the standard form $(x-h)^2 = 4p(y-k)$.

   $$(x-0)^2 = 4(y-2)$$

   $h = 0$ $\quad$ $4p = 4$, $p = 1$ $\quad$ $k = 2$

   The vertex is at (0,2) and the focus at (0,3). The equation of the directrix is $y = 1$. Since $|4p| = |4| = 4$, the ends of the latus rectum are two units to the right and two units to the left of the focus, or at (2,3) and (-2,3). The vertex (0,2), and the points (2,3) and (-2,3), can be used to sketch the parabola.

13. $$\begin{aligned} y^2-12x+24 &= 0 \\ y^2 &= 12x-24 \\ y^2 &= 12(x-2) \end{aligned}$$

   Now let's compare this equation to the standard form $(y-k)^2 = 4p(x-h)$.

   $$(y-0)^2 = 12(x-2)$$

   $k = 0$ $\quad$ $4p = 12$, $p = 3$ $\quad$ $h = 2$

   The vertex is at (2,0) and the focus at (5,0). The equation of the directrix is $x = -1$. Since $|4p| = |12| = 12$, the ends of the latus rectum are six units up and six units down from the focus, or at (5,6) and (5,-6). The vertex and the endpoints of the latus rectum can be used to sketch the parabola.

17. $x^2+6x+8y+1 = 0$

$$x^2+6x + \_ = -8y-1$$

$$x^2+6x+9 = -8y-1+9$$

$$(x+3)^2 = -8y+8 = -8(y-1)$$

Now let's compare this equation to the standard equation $(x-h)^2 = 4p(y-k)$.

$$(x-(-3))^2 = -8(y-1)$$

$h = -3$ $\quad 4p = -8$ $\quad k = 1$

$p = -2$

The vertex is at $(-3,1)$ and the focus at $(-3,-1)$. The equation of the directrix is $y = 3$. Since $|4p| = |-8| = 8$, the endpoints of the latus rectum are four units to the left and four units to the right of the focus, or at $(-7,-1)$ and $(1,-1)$. The vertex and the endpoints of the latus rectum can be used to sketch the parabola.

21. $y^2+6y-4x+1 = 0$

$$y^2+6y + \_ = 4x-1$$

$$y^2+6y+9 = 4x-1+9$$

$$(y+3)^2 = 4x+8 = 4(x+2)$$

Now let's compare this equation to the standard equation $(y-k)^2 = 4p(x-h)$.

$$(y-(-3))^2 = 4(x-(-2))$$

$k = -3$ $\quad 4p = 4$ $\quad h = -2$

$p = 1$

The vertex is at $(-2,-3)$ and the focus at $(-1,-3)$. The equation of the directrix is $x = -3$. Since $|4p| = |4| = 4$, the endpoints of the latus rectum are two units up and two units down from the focus, or at $(-1,-1)$ and $(-1,-5)$. The vertex and the endpoints of the latus rectum can be used to sketch the parabola.

25. Sketching the directrix and locating the focus helps analyze the problem.

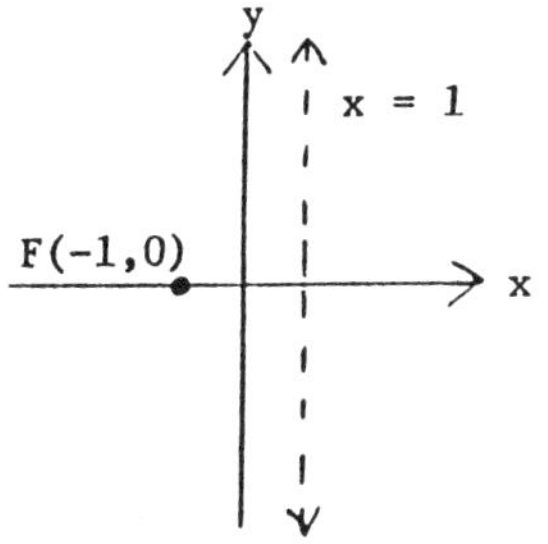

y
x = 1
F(-1,0)
x

The parabola is of the general form $y^2 = 4px$ and $p = -1$. Therefore, the equation is $y^2 = 4(-1)x = -4x$.

29. Since the directrix is a horizontal line, the equation of the parabola is of the form $(x-h)^2 = 4p(y-k)$. The vertex is half-way between the focus and directrix; so the vertex is at (3,1). This means that h = 3 and k = 1. The parabola opens upward because the focus is above the directrix and the distance between the focus and vertex is 3 units; so, p = 3. Now we can substitute 3 for h, 1 for k, and 3 for p in the general equation and simplify.

$$(x-3)^2 = 4(3)(y-1)$$

$$x^2-6x+9 = 12y-12$$

$$x^2-6x-12y+21 = 0$$

33. Let's draw a sketch of the parabola from the given information.

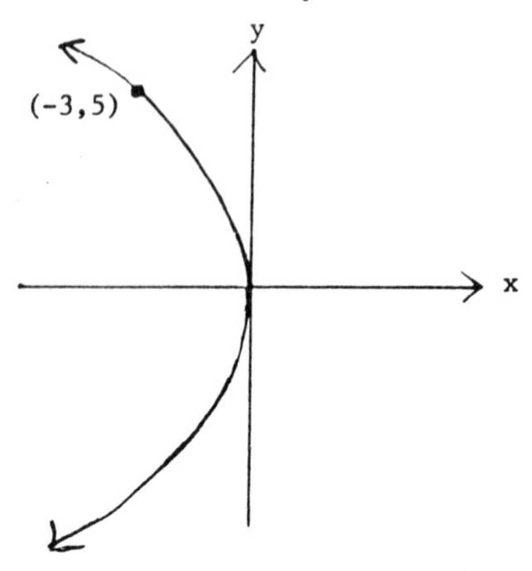

The equation is of the general form $y^2 = 4px$. Substituting -3 for x and 5 for y produces

$$5^2 = 4p(-3)$$

$$25 = -12p$$

$$-\frac{25}{12} = p.$$

Therefore, substituting $-\frac{25}{12}$ for p in $y^2 = 4px$ produces

$$y^2 = 4(-\frac{25}{12})\ x$$

$$y^2 = -\frac{25}{3}x$$

$$3y^2 = -25x$$

$$3y^2+25x = 0$$

37. The equation is of the general form $(x-h)^2 = 4p(y-k)$. We are given the vertex (7,3); so, h = 7 and k = 3. The distance between the vertex (7,3) and the focus (7,5) is 2 units and the focus is above the vertex; so p = 2. Therefore, we can substitute 7 for h, 3 for k, and 2 for p in $(x-h)^2 = 4p(y-k)$ and simplify.

$$(x-7)^2 = 4(2)(y-3)$$

$$x^2-14x+49 = 8y-24$$

$$x^2-14x-8y+73 = 0$$

41. The equation is of the general form $(x-h)^2 = 4p(y-k)$. We are given the vertex $(-9,1)$; so $h = -9$ and $k = 1$. Thus, the equation $(x-h)^2 = 4p(y-k)$ becomes $(x+9)^2 = 4p(y-1)$. Since the point $(-8,0)$ is on the graph, we can substitute -8 for x and 0 for y.

$$(-8+9)^2 = 4p(0-1)$$
$$1 = -4p$$
$$-\frac{1}{4} = p$$

Now we can substitute $-\frac{1}{4}$ for p in the equation $(x+9)^2 = 4p(y-1)$ and simplify.

$$(x+9)^2 = 4(-\frac{1}{4})(y-1)$$
$$x^2+18x+81 = -y+1$$
$$x^2+18x+y+80 = 0$$

Problem Set 9.2

1. Comparing $\frac{x^2}{4} + \frac{y^2}{1} = 1$ to $\frac{x^2}{a^2} + \frac{y^2}{b^2} = 1$ we see that $a^2 = 4$ and $b^2 = 1$. Thus, the vertices are at $(2,0)$ and $(-2,0)$, and the ends of the minor axis at $(0,1)$ and $(0,-1)$. Since $c^2 = a^2-b^2$, we obtain $c^2 = 4-1 = 3$. So the foci are at $(\sqrt{3},0)$ and $(-\sqrt{3},0)$

5. The given equation can be changed to a standard form by dividing both sides by 27.

$$\frac{9x^2+3y^2}{27} = \frac{27}{27}$$
$$\frac{x^2}{3} + \frac{y^2}{9} = 1$$

This is of the form $\frac{x^2}{b^2} + \frac{y^2}{a^2} = 1$, where $b^2 = 3$ and $a^2 = 9$. So the vertices are at $(0,3)$ and $(0,-3)$, and the ends of the minor axis are at $(\sqrt{3},0)$ and $(-\sqrt{3},0)$. Using $c^2 = a^2-b^2$, we obtain $c^2 = 9-3 = 6$. So, the foci are at $(0,\sqrt{6})$ and $(0,-\sqrt{6})$.

9. The given equation can be changed to a standard form by dividing both sides by 36.

$$\frac{12x^2+y^2}{36} = \frac{36}{36}$$
$$\frac{x^2}{3} + \frac{y^2}{36} = 1$$

This is of the form $\frac{x^2}{b^2} + \frac{y^2}{a^2} = 1$, where $b^2 = 3$ and $a^2 = 36$. So the vertices are at $(0,6)$ and $(0,-6)$, and the ends of the minor axis are at $(\sqrt{3},0)$ and $(-\sqrt{3},0)$. Using $c^2 = a^2-b^2$, we obtain $c^2 = 36-3 = 33$. So the foci are at $(0,\sqrt{33})$ and $(0,-\sqrt{33})$.

13. First, we need to change to a standard form by completing the square on both x and y.

$$4(x^2-2x+\_\_)+9(y^2-4y+\_\_) = -4$$
$$4(x^2-2x+1)+9(y^2-4y+4) = -4+4+36$$
$$4(x-1)^2 + 9(y-2)^2 = 36$$
$$\frac{(x-1)^2}{9} + \frac{(y-2)^2}{4} = 1$$

This equation is of the form $\frac{(x-h)^2}{a^2} + \frac{(y-k)^2}{b^2} = 1$, where $h = 1$, $k = 2$, $a = 3$, and $b = 2$. Thus, the vertices are three units to the right of the center of the ellipse (1,2); so the vertices are at (-2,2) and (4,2). The endpoints of the minor axis are two units up and two units down from the center; so they are at (1,4) and (1,0). Using $c^2 = a^2-b^2$, we obtain $c^2 = 9-4 = 5$. The foci are $\sqrt{5}$ units left and right of the center; so they are at $(1-\sqrt{5},2)$ and $(1+\sqrt{5},2)$.

17. First, we need to change to a standard form by completing the square on x.

$$x^2-6x+\_\_+4y^2 = -5$$
$$x^2-6x+9+4y^2 = -5+9$$
$$(x-3)^2+4(y-0)^2 = 4$$
$$\frac{(x-3)^2}{4} + \frac{(y-0)^2}{1} = 1$$

The equation is of the form $\frac{(x-h)^2}{a^2} + \frac{(y-k)^2}{b^2} = 1$, where $h = 3$, $k = 0$, $a = 2$, and $b = 1$. The vertices are two units to the left and right of the center (3,0). So the vertices are at (1,0) and (5,0). The endpoints of the minor axis are 1 unit up and down from the center; so they are at (3,1) and (3,-1). Using $c^2 = a^2-b^2$, we obtain $c^2 = 4-1 = 3$. Thus, the foci are at $(3-\sqrt{3},0)$ and $(3+\sqrt{3},0)$.

21. First, we need to change to a standard form by completing the square on both x and y.

$$2(x^2+6x+\_\_)+1(y^2-8y+\_\_) = -172$$
$$2(x^2+6x+9)+11(y^2-8y+16) = -172+18+176$$
$$2(x+3)^2+11(y-4)^2 = 22$$
$$\frac{(x+3)^2}{11} + \frac{(y-4)^2}{2} = 1$$

This equation is of the form $\frac{(x-h)^2}{a^2} + \frac{(y-k)^2}{b^2} = 1$, where $h = -3$, $k = 4$, $a = \sqrt{11}$, and $b = \sqrt{2}$. The vertices are $\sqrt{11}$ units to the left and right of the center (-3,4). So the vertices are at $(-3-\sqrt{11},4)$ and $(-3+\sqrt{11},4)$. The endpoints of the minor axis are $\sqrt{2}$ units up and down from the center; so they are at $(-3,4+\sqrt{2})$ and $(-3,4-\sqrt{2})$. Using $c^2 = a^2-b^2$, we obtain $c^2 = 11-2 = 9$. So the foci are at (-6,4) and (0,4).

25. This equation is of the form $\frac{x^2}{b^2} + \frac{y^2}{a^2} = 1$. From the given information we know that $a = 6$ and $c = 5$. Therefore, $b^2 = a^2-c^2 = 36-25 = 11$. Now substituting 36 for a^2 and 11 for b^2 produces

$$\frac{x^2}{11} + \frac{y^2}{36} = 1.$$

Multiplying both sides by 396 produces $36x^2+11y^2 = 396$.

29. From the given information we know that $c = 2$ and $b = \frac{3}{2}$. Using $a^2 = b^2+c^2$, we obtain $a^2 = \frac{9}{4} + 4 = \frac{25}{4}$. Now we can substitute $\frac{25}{4}$ for a^2 and $\frac{9}{4}$ for b^2 in the form $\frac{x^2}{b^2} + \frac{y^2}{a^2} = 1$ and simplify.

$$\frac{x^2}{\frac{9}{4}} + \frac{y^2}{\frac{25}{4}} = 1$$

$$\frac{4x^2}{9} + \frac{4y^2}{25} = 1$$

$$100x^2+36y^2 = 225$$

33. Since the vertices and foci are on the same horizontal line ($y = 1$), this ellipse has an equation of the form

$$\frac{(x-h)^2}{a^2} + \frac{(y-k)^2}{b^2} = 1.$$

The center of the ellipse is at the midpoint of the major axis. Thus,

$$h = \frac{5+(-3)}{2} = 1 \text{ and } k = 1.$$

The distance between the center (1,1) and a vertex (5,1) is 4 units; so $a = 4$. The distance between the center (1,1) and a focus (3,1) is 2; thus, $c = 2$. Using the relationship $b^2 = a^2-c^2$, we obtain $b^2 = 16-4 = 12$. Now we can substitute 1 for h, 1 for k, 16 for a^2, 12 for b^2, and simplify.

$$\frac{(x-1)^2}{16} + \frac{(y-1)^2}{12} = 1$$

$$3(x-1)^2+4(y-1)^2 = 48$$

$$3(x^2-2x+1)+4(y^2-2y+1) = 48$$

$$3x^2-6x+3+4y^2-8y+4 = 48$$

$$3x^2-6x+4y^2-8y-41 = 0$$

37. This is an ellipse of the form $\frac{x^2}{a^2} + \frac{y^2}{b^2} = 1$. The given points (2,0) and (-2,0) are the foci; so, $c = 2$. The sum of the distances is 2a; so $2a = 8$ or $a = 4$. Using $b^2 = a^2-c^2$, we obtain $b^2 = 16-4 = 12$. Substituting 16 for a^2, 12 for b^2, and simplifying produces

$$\frac{x^2}{16} + \frac{y^2}{12} = 1$$

$$3x^2+4y^2 = 48.$$

Problem Set 9.3

1. This equation is of the form $\frac{x^2}{a^2} - \frac{y^2}{b^2} = 1$, where $a^2 = 9$ and $b^2 = 4$. Hence $a = 3$ and $b = 2$. Since $a = 3$, the vertices are at $(3,0)$ and $(-3,0)$. Using $c^2 = a^2+b^2$, we obtain $c^2 = 9+4 = 13$. So the foci are at $(\sqrt{13},0)$ and $(-\sqrt{13},0)$. Then using $y = \pm \frac{b}{a}x$ for the equations of the asymptotes, we get $y = \frac{2}{3}x$ and $y = -\frac{2}{3}x$.

5. Divide both sides of the given equation by 144.

$$\frac{9y^2-16x^2}{144} = \frac{144}{144}$$

$$\frac{y^2}{16} - \frac{x^2}{9} = 1$$

This equation is of the form $\frac{y^2}{a^2} - \frac{x^2}{b^2} = 1$, where $a^2 = 16$ and $b^2 = 9$. Since $a = 4$, the vertices are at $(0,4)$ and $(0,-4)$. Using $c^2 = a^2+b^2$, we obtain $c^2 = 16+9 = 25$. So the foci are at $(0,5)$ and $(0,-5)$. Then using $y = \pm \frac{a}{b}x$ for the equations of the asymptotes, we get $y = \frac{4}{3}x$ and $y = -\frac{4}{3}x$.

9. Divide both sides of the given equation by 25.

$$\frac{5y^2-x^2}{25} = \frac{25}{25}$$

$$\frac{y^2}{5} - \frac{x^2}{25} = 1$$

This equation is of the form $\frac{y^2}{a^2} - \frac{x^2}{b^2} = 1$, where $a^2 = 5$ and $b^2 = 25$. Since $a = \sqrt{5}$, the vertices are at $(0,\sqrt{5})$ and $(0,-\sqrt{5})$. Using $c^2 = a^2+b^2$, we obtain $c^2 = 5+25 = 30$. So the foci are at $(0,\sqrt{30})$ and $(0,-\sqrt{30})$. Then using $y = \pm \frac{a}{b}x$ for the equations of the asymptotes, we get $y = \frac{\sqrt{5}}{5}x$ and $y = -\frac{\sqrt{5}}{5}x$.

13. First, let's complete the square on both x and y to help put the given equation in a standard form.

$$4(x^2-6x+\_\_)-9(y^2+2y+\_\_) = 9$$

$$4(x^2-6x+9)-9(y^2+2y+1) = 9+36-9$$

$$4(x-3)^2-9(y+1)^2 = 36$$

$$\frac{(x-3)^2}{9} - \frac{(y+1)^2}{4} = 1$$

This equation is of the form $\frac{(x-h)^2}{a^2} - \frac{(y-k)^2}{b^2} = 1$, where $a^2 = 9$ and $b^2 = 4$. Since $h = 3$ and $k = -1$, the center is at $(3,-1)$. Because $a = 3$, the vertices are 3 units to the left and 3 units to the right of the center. So the vertices are at $(0,-1)$ and $(6,-1)$. From $c^2 = a^2+b^2$, we get $c^2 = 9+4 = 13$. So the foci are $\sqrt{13}$ units to the left and right of the center. Thus, the foci are at $(3-\sqrt{13},-1)$ and $(3+\sqrt{13},-1)$. Using $a = 3$ and $b = 2$, the slopes of the asymptotes are $\frac{2}{3}$ and $-\frac{2}{3}$. Then using these slopes, the center $(3,-1)$, and the point-slope form for writing the equation of a line, the equations of the asymptotes can be determined as follows.

$$y+1 = \frac{2}{3}(x-3) \text{ and } y+1 = -\frac{2}{3}(x-3)$$

$$3y+3 = 2x-6 \text{ and } 3y+3 = -2x+6$$

$$9 = 2x-3y \text{ and } 2x+3y = 3$$

17.

$$2x^2-8x-y^2+4 = 0$$

$$2(x^2-4x+\_\_)-y^2 = -4$$

$$2(x^2-4x+4)-y^2 = -4+8$$

$$2(x-2)^2-(y-0)^2 = 4$$

$$\frac{(x-2)^2}{2} - \frac{(y-0)^2}{4} = 1$$

This equation is of the form $\frac{(x-h)^2}{a^2} - \frac{(y-k)^2}{b^2} = 1$, where $h = 2$, $k = 0$, $a^2 = 2$, and $b^2 = 4$. Thus, the center is at $(2,0)$. Because $a = \sqrt{2}$, the vertices are $\sqrt{2}$ units to the left and right of the center. So the vertices are at $(2-\sqrt{2},0)$ and $(2+\sqrt{2},0)$. From $c^2 = a^2+b^2$ we obtain $c^2 = 2+4 = 6$. Thus, the foci are at $(2-\sqrt{6},0)$ and $(2+\sqrt{6},0)$. Using $a = \sqrt{2}$ and $b = 2$, the slopes of the asymptotes are $\pm\sqrt{2}$. Using these slopes, the center $(2,0)$, and the point-slope form, the equations of the asymptotes can be derived as follows.

$$y-0 = \sqrt{2}(x-2) \quad \text{and} \quad y-0 = -\sqrt{2}(x-2)$$

$$y = \sqrt{2}\,x-2\sqrt{2} \quad \text{and} \quad y = -\sqrt{2}\,x+2\sqrt{2}$$

21.

$$x^2+4x-y^2-4y-1 = 0$$

$$x^2+4x+\_\_ -(y^2+4y+\_\_) = 1$$

$$x^2+4x+4-(y^2+4y+4) = 0$$

$$(x+2)^2-(y+2)^2 = 1$$

This equation is of the form $\frac{(x-h)^2}{a^2} - \frac{(y-k)^2}{b^2} = 1$, where $h = -2$, $k = -2$, $a^2 = 1$, and $b^2 = 1$. Thus, the center is at $(-2,-2)$. Because $a = 1$, the vertices are 1 unit to the left and right of the center. So the vertices are at $(-3,-2)$ and $(-1,-2)$. Using $c^2 = a^2+b^2$, we obtain $c^2 = 1+1 = 2$. So the foci are at $(-2-\sqrt{2},-2)$ and $(-2+\sqrt{2},-2)$. The slopes of the asymptotes are ± 1. The equations of the asymptotes can be determined using the slopes, the center of the hyperbola, and the point-slope form.

$$y+2 = 1(x+2) \text{ and } y+2 = -1(x+2)$$

$$y+2 = x+2 \quad \text{and } y+2 = -x-2$$

$$0 = x-y \quad \text{and } x+y = -4$$

25. From the given information we know that $a = 3$ and $c = 5$. Then using $b^2 = c^2-a^2$, we obtain $b^2 = 25-9 = 16$. Since the foci are on the y-axis, this equation is of the form $\frac{y^2}{a^2} - \frac{x^2}{b^2} = 1$. Substituting 9 for a^2, 16 for b^2, and simplifying produces

$$\frac{y^2}{9} - \frac{x^2}{16} = 1$$

$$16y^2-9x^2 = 144.$$

29. From the given information we know that $a = \sqrt{3}$ and $2b = 4$ or $b = 2$. Since the vertices are on the y-axis, this equation is of the form $\frac{y^2}{a^2} - \frac{x^2}{b^2} = 1$. So we can substitute 3 for a^2, 4 for b^2, and simplify.

$$\frac{y^2}{3} - \frac{x^2}{4} = 1$$

$$4y^2-3x^2 = 12$$

33. Since the vertices and foci are on the same horizontal line ($y = -3$), this hyperbola has an equation of the form

$$\frac{(x-h)^2}{a^2} - \frac{(y-k)^2}{b^2} = 1.$$

The center of the hyperbola is midway between the vertices. Thus,

$$h = \frac{6+2}{2} = 4 \text{ and } k = -3.$$

The distance between the center $(4,-3)$ and a vertex $(6,-3)$ is 2 units; so $a = 2$. The distance between the center $(4,-3)$ and a focus $(7,-3)$ is 3 units; so, $c = 3$. Then using $b^2 = c^2-a^2$, we obtain $b^2 = 9-4 = 5$. Now we can substitute 4 for h, -3 for k, 4 for a^2, 5 for b^2, and simplify.

$$\frac{(x-4)^2}{4} - \frac{(y+3)^2}{5} = 1$$

$$5(x-4)^2-4(y+3)^2 = 20$$

$$5(x^2-8x+16)-4(y^2+6y+9) = 20$$

$$5x^2-40x+80-4y^2-24y-36-20 = 0$$

$$5x^2-40x-4y^2-24y+24 = 0$$

37. Since the vertices and foci are on the same horizontal line (x-axis), the equation of this hyperbola is of the form

$$\frac{(x-h)^2}{a^2} - \frac{(y-k)^2}{b^2} = 1.$$

The center of the hyperbola is midway between the vertices. Thus,

$$h = \frac{0+4}{2} = 2 \text{ and } k = 0.$$

The distance between the center $(2,0)$ and a vertex $(0,0)$ is 2 units; so, $a = 2$. The distance between the center $(2,0)$ and a focus $(5,0)$ is 3 units; so, $c = 3$. Then using $b^2 = c^2-a^2$, we obtain $b^2 = 9-4 = 5$. Now we can substitute 2 for h, 0 for k, 4 for a^2, 5 for b^2, and simplify.

$$\frac{(x-2)^2}{4} - \frac{(y-0)^2}{5} = 1$$

$$5(x-2)^2-4(y-0)^2 = 20$$

$$5(x^2-4x+4)-4y^2-20 = 0$$

$$5x^2-20x+20-4y^2-20 = 0$$

$$5x^2-20x-4y^2 = 0$$

41. This equation is of the form $Ax+By = C$; so it is a straight line.

45. By completing the square, this equation can be put in the form

$$\frac{(x-h)^2}{a^2} - \frac{(y-k)^2}{b^2} = 1$$

So it is a hyperbola.

Problem Set 9.4

1. The first equation represents a circle and the second a straight line. By graphing these figures, we should predict one solution for the system, that is, the line should appear to be tangent to the circle.

   Rewriting the second equation as $x = 5-2y$, we can substitute $5-2y$ for x in the first equation and solve for y.

$$\begin{aligned} (5-2y)^2+y^2 &= 5 \\ 25-20y+4y^2+y^2 &= 5 \\ 5y^2-20y+20 &= 0 \\ y^2-4y+4 &= 0 \\ (y-2)^2 &= 0 \\ y &= 2 \end{aligned}$$

   If $y = 2$, then $x = 5-2(2) = 1$. So the solution set is $\{(1,2)\}$.

5. The first equation represents a circle and the second a straight line. By graphing these figures, we should see that they do not intersect. Therefore, the solution set should consist of complex numbers that are not real numbers.

   Rewriting the second equation as $x = y+4$, we can substitute $y+4$ for x in the first equation and solve for y.

$$\begin{aligned} (y+4)^2+y^2 &= 2 \\ y^2+8y+16+y^2 &= 2 \\ 2y^2+8y+14 &= 0 \\ y^2+4y+7 &= 0 \\ y &= \frac{-4 \pm \sqrt{16-28}}{2} \\ &= \frac{-4 \pm \sqrt{-12}}{2} \\ &= \frac{-4 \pm 2i\sqrt{3}}{2} \\ &= -2 \pm i\sqrt{3} \end{aligned}$$

   If $y = -2 + i\sqrt{3}$, then $x = y+4 = -2 + i\sqrt{3} + 4 = 2 + i\sqrt{3}$. If $y = -2 - i\sqrt{3}$, then $x = y+4 = -2 - i\sqrt{3} + 4 = 2 - i\sqrt{3}$. The solution set of the system is

$$\{(2 + i\sqrt{3}, -2 + i\sqrt{3}), (2 - i\sqrt{3}, -2 - i\sqrt{3})\}.$$

9. The first equation represents a line and the second a parabola. Graphing these figures should indicate one point of intersection.

Rewriting the first equation as $y = -2x-2$, we can substitute $-2x-2$ for y in the second equation and solve for x.

$$-2x-2 = x^2+4x+7$$
$$0 = x^2+6x+9$$
$$0 = (x+3)^2$$
$$-3 = x$$

If $x = -3$, then $y = -2x-2 = -3(-3)-2 = 4$.

The solution set is $\{(-3,4)\}$.

13. The first equation represents an ellipse and the second a line. Graphing these figures should indicate one point of intersection.

Rewriting the second equation as $x = 4y-9$, we can substitute $4y-9$ for x in the first equation and solve for y.

$$(4y-9)^2+2y^2 = 9$$
$$16y^2-72y+81+2y^2 = 9$$
$$18y^2-72y+72 = 0$$
$$y^2-4y+4 = 0$$
$$(y-2)^2 = 0$$
$$y = 2$$

If $y = 2$, then $x = 4(2)-9 = -1$.

The solution set is $\{(-1,2)\}$.

17. The first equation represents a line and the second a hyperbola. Graphing the two figures should indicate one point of intersection.

Rewriting the first equation as $x = y+2$, we can substitute $y+2$ for x in the second equation and solve for y.

$$(y+2)^2-y^2 = 16$$
$$y^2+4y+4-y^2 = 16$$
$$4y = 12$$
$$y = 3$$

If $y = 3$, then $x = y+2 = 3+2 = 5$.

The solution set is $\{(5,3)\}$.

21. Both equations represent parabolas. By graphing the two parabolas, we should predict one point of intersection.

Equating the two expressions for y and solving for x produces the following.

$$x^2+2x-1 = x^2+4x+5$$
$$-6 = 2x$$
$$-3 = x$$

If $x = -3$, then $y = x^2+2x-1 = (-3)^2+2(-3)-1 = 2$.

The solution set is $\{(-3,2)\}$.

25. Both equations represent hyperbolas and graphing them should indicate four points of intersection.

To solve the system, let's use the elimination method.

$$\begin{array}{lll} -9x^2+8y^2 = 6 & \xrightarrow{\text{Multiply by 3.}} & -27x^2+24y^2 = 18 \\ 8x^2-3y^2 = 7 & \xrightarrow{\text{Multiply by 8.}} & \underline{64x^2-24y^2 = 56} \\ & & 37x^2 = 74 \\ & & x^2 = 2 \\ & & x = \pm\sqrt{2} \end{array}$$

If $x = \sqrt{2}$, then $8y^2-9x^2 = 6$ becomes

$$\begin{aligned} 8y^2-9(\sqrt{2})^2 &= 6 \\ 8y^2-18 &= 6 \\ 8y^2 &= 24 \\ y^2 &= 3 \\ y &= \pm\sqrt{3}. \end{aligned}$$

If $x = -\sqrt{2}$, then $8y^2-9x^2 = 6$ becomes

$$\begin{aligned} 8y^2-9(-\sqrt{2})^2 &= 6 \\ 8y^2-18 &= 6 \\ 8y^2 &= 24 \\ y^2 &= 3 \\ y &= \pm\sqrt{3}. \end{aligned}$$

The solution set is $\{(\sqrt{2},\sqrt{3}),(\sqrt{2},-\sqrt{3}),(-\sqrt{2},\sqrt{3}),(-\sqrt{2},-\sqrt{3})\}$.

29. The first equation represents a hyperbola and the second equation represents a line. By graphing them we should be able to predict two points of intersection.

The first equation can be written as $y = \frac{3}{x}$. Then we can substitute $\frac{3}{x}$ for y in the second equation.

$$\begin{aligned} 2x+2y &= 7 \\ 2x+2\left(\frac{3}{x}\right) &= 7,\ x \neq 0 \\ 2x + \frac{6}{x} &= 7 \\ 2x^2+6 &= 7 \\ 2x^2-7x+6 &= 0 \\ (2x-3)(x-2) &= 0 \end{aligned}$$

$$2x-3 = 0 \text{ or } x-2 = 0$$

$$x = \frac{3}{2} \text{ or } \quad x = 2$$

If $x = \frac{3}{2}$, then $y = \frac{3}{x} = \frac{3}{\frac{3}{2}} = 2.$

If $x = 2$, then $y = \frac{3}{x} = \frac{3}{2}.$

The solution set is $\{(\frac{3}{2},2),(2,\frac{3}{2})\}$.

33. We can equate the two values for y.

$$e^x - 1 = 2e^{-x}$$
$$e^x - 2e^{-x} - 1 = 0$$
$$e^x - \frac{2}{e^x} - 1 = 0$$
$$e^{2x} - 2 - e^x = 0$$
$$e^{2x} - e^x - 2 = 0$$
$$(e^x - 2)(e^x + 1) = 0$$
$$e^x - 2 = 0 \text{ or } e^x + 1 = 0$$
$$e^x = 2 \text{ or } e^x = -1$$

The equation $e^x = -1$ has no real solutions. The equation $e^x = 2$ can be solved as follows.

$$e^x = 2$$
$$\ell n\, e^x = \ell n\, 2$$
$$x \ell n\, e = \ell n\, 2$$
$$x = \ell n\, 2$$

If $x = \ell n\, 2$, then $y = e^x - 1 = e^{\ell n\, 2} - 1 = 2 - 1 = 1$.

The solution set is $\{(\ell n\, 2, 1)\}$.

CHAPTER 10

Problem Set 10.1

1. $$a_n = 3n-7$$

$a_1 = 3(1)-7 = -4$; $a_2 = 3(2)-7 = -1$; $a_3 = 3(3)-7 = 2$;

$a_4 = 3(4)-7 = 5$; $a_5 = 3(5)-7 = 8$

5. $$a_n = 3n^2-1$$

$a_1 = 3(1)^2-1 = 2$; $a_2 = 3(2)^2-1 = 11$; $a_3 = 3(3)^2-1 = 26$;

$a_4 = 3(4)^2-1 = 47$; $a_5 = 3(5)^2-1 = 74$

9. $$a_n = 2^{n+1}$$

$a_1 = 2^{1+1} = 2^2 = 4$; $a_2 = 2^{2+1} = 2^3 = 8$; $a_3 = 2^{3+1} = 2^4 = 16$;

$a_4 = 2^{4+1} = 2^5 = 32$; $a_5 = 2^{5+1} = 2^6 = 64$

13. $a_{25} = (-1)^{25+1} = (-1)^{26} = 1$ $\qquad a_{50} = (-1)^{50+1} = (-1)^{51} = -1$

17. The common difference, d, is $-1-2 = -3$ and the first term a_1, is 2. Substitute these values into $a_n = a_1+(n-1)d$ and simplify.

$$\begin{aligned} a_n &= 2+(n-1)(-3) \\ &= 2-3n+3 \\ &= -3n+5 \end{aligned}$$

21. The common difference, d, is $6-2 = 4$ and the first term, a_1, is 2. Substitute these values into $a_n = a_1+(n-1)d$ and simplify.

$$\begin{aligned} a_n &= 2+(n-1)(4) \\ &= 2+4n-4 \\ &= 4n-2 \end{aligned}$$

25. Use $a_n = a_1+(n-1)d$.

$$\begin{aligned} a_{15} &= 3+(15-1)(5) \\ &= 3+(14)(5) \\ &= 73 \end{aligned}$$

29. Use $a_n = a_1+(n-1)d$.

$$\begin{aligned} a_{52} &= 1+(52-1)(\tfrac{2}{3}) \\ &= 1+(51)(\tfrac{2}{3}) \\ &= 35 \end{aligned}$$

33. Use $a_n = a_1+(n-1)d$ with $a_3 = 20$ (the third term is 20) and $a_7 = 32$ (the seventh term is 32.)

$$20 = a_1+(3-1)d = a_1+2d$$

$$32 = a_1+(7-1)d = a_1+6d$$

Now we can solve the system

$$\begin{pmatrix} a_1+2d = 20 \\ a_1+6d = 32 \end{pmatrix}$$

which produces $a_1 = 14$ and $d = 3$. Therefore,

$$a_{25} = 14+(25-1)3 = 14+24(3) = 86.$$

37. First, use $a_n = a_1+(n-1)d$ to find the 40th term.

$$a_{40} = 2+(40-1)4 = 2+39(4) = 158$$

Now use the sum formula $S_n = \dfrac{n(a_1+a_n)}{2}$.

$$S_{40} = \frac{40(2+158)}{2} = 3200$$

41. First, use $a_n = a_1+(n-1)d$ to find the 50th term.

$$a_{50} = \frac{1}{2}+(50-1)(\frac{1}{2}) = \frac{1}{2}+49(\frac{1}{2}) = 25$$

Now use the sum formula.

$$S_{50} = \frac{50(\frac{1}{2}+25)}{2} = 637.5$$

45. Use $a_n = a_1+(n-1)d$ to find the number of terms.

$$146 = 2+(n-1)6$$
$$146 = 2+6n-6$$
$$150 = 6n$$
$$25 = n$$

Then use the sum formula.

$$S_{25} = \frac{25(2+146)}{2} = 1850$$

49. Use $a_n = a_1+(n-1)d$ to find the number of terms.

$$119 = -5+(n-1)2$$
$$119 = -5+2n-2$$
$$126 = 2n$$
$$63 = n$$

Now use the sum formula.

$$S_{63} = \frac{63(-5+119)}{2} = 3591$$

53. We need to find the following sum.

$18+20+22+...+482$

First, let's use $a_n = a_1+(n-1)d$ to find the number of terms.

$$482 = 18+(n-1)2$$
$$482 = 18+2n-2$$
$$466 = 2n$$
$$233 = n$$

Now we can use the sum formula.

$$S_{233} = \frac{233(18+482)}{2} = 58,250$$

57. First, use $a_n = -4n-1$ to find the 1st and 25th terms.

$$a_1 = -4(1)-1 = -5$$
$$a_{25} = -4(25)-1 = -101$$

Now use the sum formula.

$$S_{25} = \frac{25(-5+(-101))}{2} = -1325$$

61. $a_1 = -2(1)+4 = 2$ and $a_{30} = -2(30)+4 = -56$

Now use the sum formula.

$$S_{30} = \frac{30(2+(-56))}{2} = -810$$

65. $a_{10} = 4(10) = 40$ and $a_{20} = 4(20) = 80$

Now use the sum formula.

$$S_{11} = \frac{11(40+80)}{2} = 660$$

69. By letting i take on successive values from 3 through 8, inclusive, we generate the following series.

$21+36+55+78+105+136$

Notice that it is not an arithmetic series. Therefore, we have no formula and we will simply add the numbers to get a sum of 431.

73. $a_1 = 3(1)+1 = 4$; $a_2 = 3(2)+1 = 7$; $a_3 = 3(3)+1 = 10$;

$a_4 = 4(4)-3 = 13$; $a_5 = 4(5)-3 = 17$; $a_6 = 4(6)-3 = 21$

77. $a_1 = 1$; $a_2 = 1$; $a_3 = 1+1 = 2$;

$a_4 = 1+2 = 3$; $a_5 = 2+3 = 5$; $a_6 = 3+5 = 8$

Problem Set 10.2

1. $a_n = a_1 r^{n-1} = 3(2)^{n-1}$

5. $a_n = a_1 r^{n-1} = (\frac{1}{4})(\frac{1}{2})^{n-1} = \frac{1}{2^2 \cdot 2^{n-1}} = \frac{1}{2^{n+1}} = (\frac{1}{2})^{n+1}$

9. $a_n = a_1 r^{n-1} = 1(.3)^{n-1} = (.3)^{n-1}$

13. $a_8 = \frac{1}{2}(2)^{8-1} = \frac{1}{2}(2)^7 = \frac{2^7}{2} = 2^6 = 64$

17. $a_{10} = (1)(-2)^{10-1} = (-2)^9 = -512$

21. Use $a_n = a_1 r^{n-1}$ to represent the 5th term.

$$a_5 = a_1(2)^4 = \frac{32}{3}$$

$$16a_1 = \frac{32}{3}$$

$$a_1 = \frac{32}{3} \cdot \frac{1}{16} = \frac{2}{3}$$

25. Remember that for such problems you can either use the sum formula or the technique illustrated in the text. Using the sum formula, your work could take on the following format.

$$S_n = \frac{a_1 r^n - a_1}{r-1}$$

$$S_{10} = \frac{(1)(2)^{10}-1}{2-1} = \frac{2^{10}-1}{1} = 1023$$

29. $$S_8 = \frac{8(\frac{3}{2})^8 - 8}{\frac{3}{2} - 1} = \frac{8\left((\frac{3}{2})^8 - 1\right)}{\frac{1}{2}} = 16\left((\frac{3}{2})^8 - 1\right) = 16(\frac{6305}{256})$$

$$= \frac{6305}{16} = 394\frac{1}{16}$$

33. Let's use the technique illustrated in the text rather than the sum formula.

$$S = 9+27+81+\ldots+729 \quad (1)$$
$$3S = 27+81+\ldots+729+2187 \quad (2)$$
$$2S = 2187-9 \quad \leftarrow \text{Equation (2) minus equation (1)}$$
$$2S = 2178$$
$$S = 1089$$

37.
$$S = -1+3+(-9)+\ldots+(-729) \quad (1)$$
$$3S = (-3)+9+\ldots+(729)+(-2187) \quad (2)$$
$$4S = (-1)+(-2187) \quad \leftarrow \text{Equation (1) plus equation (2)}$$
$$4S = -2188$$
$$S = -547$$

41. Generate the series by letting i take on successive values from 2 through 5, inclusive.

$$-27+81+(-243)+729$$

This sum is 540.

45. Use the sum formula $S_\infty = \frac{a_1}{1-r}$ with $a_1 = 2$ and $r = \frac{1}{2}$.

$$S_\infty = \frac{2}{1-\frac{1}{2}} = \frac{2}{\frac{1}{2}} = 4$$

49. Since $r = 2$, this infinite sequence has no sum.

53. Use $S_\infty = \frac{a_1}{1-r}$ with $a_1 = \frac{1}{2}$ and $r = \frac{3}{4}$.

$$S_\infty = \frac{\frac{1}{2}}{1-\frac{3}{4}} = \frac{\frac{1}{2}}{\frac{1}{4}} = (\frac{1}{2})(\frac{4}{1}) = 2$$

57. The repeating decimal $.\overline{3}$ can be written as the infinite geometric series

$$.3+.03+.003+\ldots,$$

where $a_1 = .3$ and $r = .1$.

$$S_\infty = \frac{.3}{1-.1} = \frac{.3}{.9} = \frac{3}{9} = \frac{1}{3}$$

61. The repeating decimal $.\overline{123}$ can be written as the infinite geometric series

$$.123+.000123+.000000123+\ldots,$$

where $a_1 = .123$ and $r = .001$.

$$S_\infty = \frac{.123}{1-.001} = \frac{.123}{.999} = \frac{123}{999} = \frac{41}{333}$$

65. The repeating decimal $.2\overline{14}$ can be written as

$$[.2]+[.014+.00014+.0000014+\ldots]$$

where

$$.014+.00014+.0000014+\ldots$$

is an infinite geometric series with $a_1 = .014$ and $r = .01$.

$$S_\infty = \frac{.014}{1-.01} = \frac{.014}{.99} = \frac{14}{990} = \frac{7}{495}$$

Now we can add .2 and $\frac{7}{495}$.

$$.2\overline{14} = .2 + \frac{7}{495}$$

$$= \frac{1}{5} + \frac{7}{495}$$

$$= \frac{99+7}{495}$$

$$= \frac{106}{495}$$

Problem Set 10.3

1. The following sequence represents his annual salary beginning in 1960.

   9500,10200,10900,...

   This is an arithmetic sequence with $a_1 = 9500$ and $d = 700$. We are looking for the 22nd term.

   $$a_{22} = 9500+(22-1)(700)$$
   $$= 9500+21(700) = 24200$$

   His salary for 1981 was $24,200.

5. Each near's enrollment is 110% of the previous year's enrollment. So we have a geometric sequence with $a_1 = 5000$ and $r = 1.1$. We are looking for the 5th term.

   $a_5 = 5000(1.1)^4 = 7320$, rounded to the nearest whole number

9. If $\frac{1}{3}$ of the water is taken out each day, then $\frac{2}{3}$ of the water remains. Thus, we can consider a geometric sequence with $a_1 = \frac{2}{3}(5832) = 3888$ and $r = \frac{2}{3}$. We are looking for the 6th term.

   $a_6 = 3888(\frac{2}{3})^5 = 512$ gallons

13. The geometric sequence 1,2,4,8,... represents the savings, in cents, on a day-by-day basis. On the 15th day, the savings will be

    $a_{15} = 1(2)^{14} = 16384$ or $163.84.

    The total savings for the 15 days will be

    $S_{15} = \frac{1(2)^{15}-1}{2-1} = 32767$ or $327.67.

17. The sequence 16,48,80,112,... is an arithmetic sequence with a common difference of 32. In the 11th second, the distance it falls is

    $a_{11} = 16+(11-1)32 = 336$ feet.

    The total distance it falls is

    $S_{11} = \frac{11(16+336)}{2} = 1936$ feet.

21. The geometric sequence

    1458,972,648,...

    where $r = \frac{2}{3}$ represents the bouncing ball. The first term, 1458, is the original distance that the ball falls. The second term, 972, represents the first bounce up and back down. Subsequent terms represent subsequent bounces up and down. Therefore, using the sum formula we obtain

    $$S_6 = \frac{a_1-a_1r^6}{1-r} = \frac{1458-1458(\frac{2}{3})^6}{\frac{1}{3}} = 3990 \text{ feet.}$$

25. If $\frac{1}{3}$ of the air is pumped out on each stroke, then $\frac{2}{3}$ of the air remains. Therefore, the geometric sequence

$$\frac{2}{3}, \frac{4}{9}, \frac{8}{27}, \ldots,$$

with $r = \frac{2}{3}$ represents the air remaining after each stroke.

$$a_7 = \frac{2}{3}\left(\frac{2}{3}\right)^6 = \left(\frac{2}{3}\right)^7 = \frac{128}{2187} = 5.9\%$$

Problem Set 10.4

1. Part 1: If $n = 1$, then $a_1 = 1$ and $S_1 = \frac{1(1+1)}{2} = 1$. So the formula for S_n holds for $n = 1$.

   Part 2: We need to prove that if $S_k = \frac{k(k+1)}{2}$, then $S_{k+1} = \frac{(k+1)(k+2)}{2}$.

   Proof:

$$\begin{aligned} S_{k+1} &= S_k + a_{k+1} \\ &= \frac{k(k+1)}{2} + k+1 \\ &= \frac{k(k+1)+2(k+1)}{2} \\ &= \frac{(k+1)(k+2)}{2} \end{aligned}$$

5. Part 1: If $n = 1$, then $a_1 = 2^1 = 2$ and $S_1 = 2(2^1-1) = 2(1) = 2$. So the formula for S_n holds for $n = 1$.

   Part 2: We need to prove that if $S_k = 2(2^k-1)$, then $S_{k+1} = 2(2^{k+1}-1)$.

   Proof:

$$\begin{aligned} S_{k+1} &= S_k + a_{k+1} \\ &= 2(2^k-1)+2^{k+1} \\ &= 2^{k+1}-2+2^{k+1} \\ &= 2\cdot 2^{k+1}-2 \\ &= 2(2^{k+1}-1) \end{aligned}$$

9. Part 1: If $n = 1$, then $a_1 = \frac{1}{1(1+1)} = \frac{1}{2}$ and $S_1 = \frac{1}{1+1} = \frac{1}{2}$. So the formula for S_n holds for $n = 1$.

   Part 2: We need to prove that if $S_k = \frac{k}{k+1}$, then $S_{k+1} = \frac{k+1}{k+2}$.

Proof:

$$S_{k+1} = S_k + a_{k+1}$$

$$= \frac{k}{k+1} + \frac{1}{(k+1)(k+2)}$$

$$= \frac{k(k+2)+1}{(k+1)(k+2)}$$

$$= \frac{k^2+2k+1}{(k+1)(k+2)}$$

$$= \frac{(k+1)(k+1)}{(k+1)(k+2)}$$

$$= \frac{k+1}{k+2}$$

13. Part 1: If $n = 1$, then $n^2 \geq n$ becomes $1^2 \geq 1$ which is a true statement.

Part 2: We need to prove that if $k^2 \geq k$, then $(k+1)^2 \geq k+1$.

Proof:

$$k^2 \geq k$$

$$k^2+2k+1 \geq k+2k+1 \quad \text{Add } 2k+1 \text{ to both sides.}$$

$$(k+1)^2 \geq 3k+1$$

Furthermore,

$$k \geq 0$$

$$2k \geq 0$$

$$2k+k+1 \geq k+1 \quad \text{Add } k+1 \text{ to both sides.}$$

$$3k+1 \geq k+1$$

Therefore, since $(k+1)^2 \geq 3k+1$ and $3k+1 \geq k+1$, by transitivity we know that $(k+1)^2 \geq k+1$, which is what we were trying to prove.

17. Part 1: If $n = 1$, then 6^n-1 becomes $6^1-1 = 5$ which is divisible by 5.

Part 2: We need to prove that if 6^k-1 is divisible by 5, then $6^{k+1}-1$ is divisible by 5.

Proof:

If 6^k-1 is divisible by 5, then for some integer x we have $6^k-1 = 5x$. Therefore,

$$6^k-1 = 5x$$

$$6^k = 1+5x$$

$$6(6^k) = 6(1+5x) \quad \text{Multiply both sides by 6.}$$

$$6^{k+1} = 6+30x$$

$$6^{k+1} = 1+5+30x$$

$$6^{k+1}-1 = 5(1+6x)$$

Therefore, $6^{k+1}-1$ is divisible by 5.

CHAPTER 11

Problem Set 11.1

1. Task 1: Choose a skirt, for which there are 2 choices.

   Task 2: Choose a blouse, for which there are 10 choices.

   Task 1 followed by Task 2 can be done in $2 \cdot 10 = 20$ ways.

5. Task 1: Choose the units digit. It must be even, so there are 4 choices.

   Task 2: Choose the tens digit. Since one digit was used in Task 1, there are now $8-1 = 7$ choices.

   Task 3: Choose the hundreds digit. Since one digit was used for each of the previous tasks, there are now $8-2 = 6$ choices.

   Task 1 followed by Task 2 followed by Task 3 can be done in $4 \cdot 7 \cdot 6 = 168$ ways.

9. Think in terms of a tree diagram. Each of the 2 sex classifications branches into 3 party affiliation classifications, which in turn each branch into 6 family income classifications. Therefore, there are $(2)(3)(6) = 36$ different combined classifications possible.

13. The tunes can be played in $6 \cdot 5 \cdot 4 \cdot 3 \cdot 2 \cdot 1 = 720$ different orderings.

17. Task 1: Choose the left-end seat. There are 3 choices since neither Amy nor Bob is to occupy an end seat.

    Task 2: Choose the right-end seat. There are now 2 choices left for that seat.

    Task 3: Fill the remaining three seats. This can be done in $3 \cdot 2 \cdot 1 = 6$ ways.

    Therefore, there are $(3)(2)(6) = 36$ ways of seating the people.

21. Each letter can be dropped in any one of 3 mailboxes. Therefore, there are

    $$3 \cdot 3 \cdot 3 \cdot 3 \cdot 3 = 243$$

    ways of mailing the letters.

25. Each die can fall in any one of 6 ways. Therefore, there are

    $$6 \cdot 6 \cdot 6 = 216$$

    possible outcomes.

29. All 4-digit numbers are greater than 400, so there are

    $$4 \cdot 3 \cdot 2 \cdot 1 = 24$$

    4-digit numbers.

    To count the 3-digit numbers that are greater than 400, let's think as follows:

Task 1: Choose the hundreds digit. There are 2 choices (the 4 or the 5).

Task 2: Choose the tens digit. There are now 4-1 = 3 choices.

Task 3: Choose the units digit. There are now 4-2 = 2 choices.

So there are $2\cdot3\cdot2 = 12$ 3-digit numbers and, therefore, a total of 24+12 = 36 numbers greater than 400.

33. There are 2 choices for each question; therefore, there are

$$2^{10} = 1024$$

total possibilities.

37. (a) There are $26\cdot26\cdot9\cdot10\cdot10\cdot10 = 6{,}084{,}000$ different possibilities.

(c) There are $26\cdot26\cdot9\cdot9\cdot8\cdot7 = 3{,}066{,}336$ different possibilities.

Problem Set 11.2

1. $P(5,3) = 5\cdot4\cdot3 = 60$

5. $C(7,2) = \frac{P(7,2)}{2!} = \frac{7\cdot6}{2} = 21$

9. $C(15,2) = \frac{P(15,2)}{2!} = \frac{15\cdot14}{2} = 105$

13. $P(4,4) = 4\cdot3\cdot2\cdot1 = 24$

17. (a) $P(8,3) = 8\cdot7\cdot6 = 336$

21. There are $C(7,4) = \frac{P(7,4)}{4!} = \frac{7\cdot6\cdot5\cdot4}{4\cdot3\cdot2\cdot1} = 35$ ways of choosing the four women.

There are $C(8,4) = \frac{P(8,4)}{4!} = \frac{8\cdot7\cdot6\cdot5}{4\cdot3\cdot2\cdot1} = 70$ ways of choosing the four men.

Therefore, there are $35\cdot70 = 2450$ committees that can be formed.

25. If A is to be in the subsets, then that leaves two letters to be chosen from {B,C,D,E,F}.

$$C(5,2) = \frac{P(5,2)}{2!} = \frac{5\cdot4}{2} = 10$$

29. $\frac{9!}{3!4!2!} = \frac{9\cdot8\cdot7\cdot6\cdot5\cdot4\cdot3\cdot2\cdot1}{3\cdot2\cdot1\cdot4\cdot3\cdot2\cdot1\cdot2\cdot1} = 1260$

33. $\frac{6!}{4!2!} = \frac{6\cdot5\cdot4\cdot3\cdot2\cdot1}{4\cdot3\cdot2\cdot1\cdot2\cdot1} = 15$

37. There are $C(9,2) = 36$ ways of getting two good bulbs from the 9 good ones, and $C(4,1) = 4$ ways of getting the 1 defective from the four defective bulbs. So there are $36\cdot4 = 144$ samples containing two good and one defective bulb.

To find how many samples will contain at least one defective bulb, let's subtract the number of samples that contain all good bulbs from the total number of samples.

$$C(13,3)-C(9,3) = 286-84 = 202$$

41. The desired subsets contain A and three other letters chosen from {C,D,E,F,G} or B and three other letters chosen from {C,D,E,F,G}.

$$C(5,3)+C(5,3) = 10+10 = 20$$

45. $5\cdot4\cdot3\cdot2\cdot1 = 120$

Problem Set 11.3

1. Let $S = \{(H,H),(H,T),(T,H),(T,T)\}$ where $n(S) = 4$. Let $E = \{(H,T),(T,H)\}$ where $n(E) = 2$. Therefore,

$$P(E) = \frac{2}{4} = \frac{1}{2}.$$

5. If three coins are tossed, there are $2 \cdot 2 \cdot 2 = 8$ possible outcomes; so, $n(S) = 8$. Let $E = \{(H,H,H)\}$, where $n(E) = 1$. Thus,

$$P(E) = \frac{1}{8}.$$

9. If 4 coins are tossed, there are $2 \cdot 2 \cdot 2 \cdot 2 = 16$ possible outcomes; so, $n(S) = 16$. Let $E = \{(H,H,H,H)\}$, where $n(E) = 1$. Therefore,

$$P(E) = \frac{1}{16}.$$

13. If one die is tossed, there are 6 possible outcomes; so, $n(S) = 6$. Let $E = \{3,6\}$, where $n(E) = 2$. Therefore,

$$P(E) = \frac{2}{6} = \frac{1}{3}.$$

17. There are $6 \cdot 6 = 36$ possible outcomes for tossing two dice. So $n(S) = 36$. Let $E = \{(1,5),(5,1),(2,4),(4,2),(3,3)\}$, where $n(E) = 5$. Therefore,

$$P(E) = \frac{5}{36}.$$

21. As with Problem 17, $n(S) = 36$. Let $E = \{(1,4),(2,4),(3,4),(4,4),(5,4),(6,4),(4,1),(4,2),(4,3),(4,5),(4,6)\}$, where $n(E) = 11$. Therefore,

$$P(E) = \frac{11}{36}.$$

25. There are 52 possible outcomes; so, $n(S) = 52$. There are 13 spades and 13 diamonds; so, $n(E) = 13+13 = 26$. Therefore,

$$P(E) = \frac{26}{52} = \frac{1}{2}.$$

29. There are 25 possible outcomes; so, $n(S) = 25$. Let $E = \{(2,3,5,7,11,13,17,19,23\}$, where $n(E) = 9$. Therefore,

$$P(E) = \frac{9}{25}.$$

33. The sample space consists of all 2-boy committees that can be chosen from five boys. Therefore, $n(S) = C(5,2) = 10$.

The event space consists of all 2-boy committees that do not contain both Bill and Carl. So the only committee not wanted is {Bill,Carl}. Thus, the event space has $10-1 = 9$ elements, that is, $n(E) = 9$. Therefore,

$$P(E) = \frac{9}{10}.$$

37. The sample space consists of all 5-person committees that can be formed from eight people. Therefore, $n(S) = C(8,5) = 56$.

The event space consists of 5-person committees that contain Chad and four others chosen from {Barb, Dawn, Eric, Fern, George, Harriet} or that contain Dawn and four others chosen from the same group. So, $n(E) = C(6,4)+C(6,4) = 15+15 = 30$.

Therefore, $P(E) = \frac{30}{56} = \frac{15}{28}$.

41. The sample space consists of all possible 3-item samples that can be chosen from 10 items. Thus, $n(S) = C(10,3) = 120$.

The event space consists of 3-item samples having 2 defective and 1 non-defective. The 2 defective must come from a total of 2 defective and the 1 non-defective from a total of 8 non-defective. So $n(E) = C(2,2) \cdot C(8,1) = 8$.

Therefore, $P(E) = \frac{8}{120} = \frac{1}{15}$.

45. The sample space consists of all possible 4-girl committees that can be chosen from 5 girls. Thus, $n(S) = C(5,4) = 5$.

The event space consists of all possible 4-girl committees that do not contain Elaine. Obviously, that is only one such committee, so $n(E) = 1$.

Therefore, $P(E) = \frac{1}{5}$.

49. The number of 4-digit numbers that can be formed is $4 \cdot 3 \cdot 2 \cdot 1 = 24$; so, $n(S) = 24$.

To count the numbers greater than 4000, think in terms of tasks.

Task 1: Choose the thousands digit; there are $\underline{2}$ choices (4 or 6).

Task 2: Choose the other three digits in $3 \cdot 2 \cdot 1 = \underline{6}$ different ways.

Thus, there are $2 \cdot 6 = 12$ different 4-digit numbers greater than 4000; so, $n(E) = 12$.

Therefore, $P(E) = \frac{12}{24} = \frac{1}{2}$.

53. The ten men can be divided into two 5-man teams in $\frac{C(10,5)}{2} = 126$ ways; so, $n(S) = 126$.

A team consisting of Al, Bob, Carl, and two others can be formed in $C(7,2) = 21$ ways; so, $n(E) = 21$.

Therefore, $P(E) = \frac{21}{126} = \frac{1}{6}$.

57. If five coins are tossed, there are $2^5 = 32$ possible outcomes; so, $n(S) = 32$.

The event of "not getting more than 3 heads" is made up of 3 heads and 2 tails, or 2 heads and 3 tails, or 1 head and 4 tails, or 5 tails. Thus, $n(E) = C(5,3)+C(5,2)+C(5,1)+C(5,5) = 10+10+5+1 = 26$.

Therefore, $P(E) = \frac{26}{32} = \frac{13}{16}$.

61. In each suit the number of straight flushes can be listed and counted. For example, in hearts the straight flushes are A2345, 23456, 34567, 45678, 56789, 678910, 78910J, 8910JQ, 910JQK, and 10JQKA. So there are 10 in each suit, or a total of 4(10) = 40 straight flushes.

65. Let's determine all possible straights by thinking in terms of tasks.

Task 1: Choose the first card of the straight. Since a straight cannot begin with a jack, queen, or a king, there are 52-12 = 40 choices for the first card.

Task 2: Choose the remaining 4 cards. For each of these cards there are 4 choices. Thus, the remaining 4 cards can be chosen in $4 \cdot 4 \cdot 4 \cdot 4 = 256$ ways.

So, altogether there are $40 \cdot 256 = 10{,}240$ straights, but this includes the straight flushes, of which there are 40 as determined in Problem 61. Therefore, there are $10{,}240 - 40 = 10{,}200$ straights that are not straight flushes.

69. This number can be determined by subtracting the total of Problems 61-68, inclusive, from the total number of hands, 2,598,960.

Problem Set 11.4

1. If two dice are tossed, there are $6 \cdot 6 = 36$ possible outcomes; so, $n(S) = 36$.

Let $E = \{(1,5),(5,1),(2,4),(4,2),(3,3)\}$, where $n(E) = 5$.

Therefore, $P(E) = \frac{5}{36}$.

5. If three dice are tossed, there are $6 \cdot 6 \cdot 6 = 216$ possible outcomes; so, $n(S) = 216$.

Let $E = \{(1,1,1)\}$, where $n(E) = 1$.

Therefore, $P(E) = \frac{1}{216}$.

9. If four coins are tossed, there are $2 \cdot 2 \cdot 2 \cdot 2 = 16$ possible outcomes; so, $n(S) = 16$.

Let $E = \{(H,H,H,H)\}$, where $n(E) = 1$.

Therefore, $P(E) = \frac{1}{16}$.

13. If five coins are tossed, there are $2 \cdot 2 \cdot 2 \cdot 2 \cdot 2 = 32$ possible outcomes; so, $n(S) = 32$.

Let $E = \{(T,T,T,T,T)\}$, where $n(E) = 1$.

Therefore, $P(E) = \frac{1}{32}$.

17. Let S be the familiar sample space of ordered pairs for this type of problem where $n(S) = 36$. Let E be the event of not getting a double. Then E' is the complementary event of getting a double and $E' = \{(1,1),(2,2),(3,3),(4,4),(5,5),(6,6)\}$, where $n(E') = 6$. Therefore, $P(E') = \frac{6}{36} = \frac{1}{6}$ and from this we can determine P(E).

$P(E) = 1 - P(E') = 1 - \frac{1}{6} = \frac{5}{6}$.

21. A subset of 2 letters can be chosen from the set of 9 letters in $C(9,2) = 36$ ways. Thus, $n(S) = 36$.

Let E be the event that the subset contains at least one vowel. Then E' is the complementary event that the subset contains no vowels. Thus, $n(E') = C(6,2) = 15$. Therefore, $P(E') = \frac{15}{36} = \frac{5}{12}$, and $P(E) = 1 - P(E') = 1 - \frac{5}{12} = \frac{7}{12}$.

25. If one die is tossed, there are 6 possible outcomes; so, $n(S) = 6$.

Let $E = \{2\}$ be the event a "2."

Let $F = \{1,3,5\}$ be the event "an odd number."

Then $E \cap F = \emptyset$ and by Property 11.3 we get

$$p(E \cup F) = \frac{1}{6} + \frac{3}{6} - \frac{0}{6} = \frac{4}{6} = \frac{2}{3}.$$

29. If two dice are tossed, there are $6 \cdot 6 = 36$ possible outcomes; so, $n(S) = 36$.

Let $E = \{(4,6),(6,4),(5,5)\}$ be the event "a sum of 10." Let $F = \{(3,6),(6,3),(5,4),(4,5),(4,6),(6,4),(5,5),(6,5),(5,6),(6,6)\}$ be the event "a sum greater than eight." Then $E \cap F = \{(4,6),(6,4),(5,5)\}$ and by Property 11.3 we obtain

$$P(E \cup F) = \frac{3}{36} + \frac{10}{36} - \frac{3}{36} = \frac{10}{36} = \frac{5}{18}.$$

33. If three coins are tossed, there are $2 \cdot 2 \cdot 2 = 8$ possible outcomes; so, $n(s) = 8$.

Let $E = \{(H,H,T),H,T,H),(T,H,H),(H,H,H)\}$ be the event "at least two heads." Let $F = \{(H,H,T),(H,T,H),(T,H,H)\}$ be the event "exactly one tail." Then $E \cap F = \{(H,H,T),(H,T,H),(T,H,H)\}$ and by Property 11.3 we obtain

$$P(E \cup F) = \frac{4}{8} + \frac{3}{8} - \frac{3}{8} = \frac{4}{8} = \frac{1}{2}.$$

37. Each entry in the table can be converted to a probability statement by dividing by 1000, the total number of drivers. Then Property 11.3 can be applied.

(a) $P(A \cup R) = .06 + .395 - .045 = .41$

(c) $P(A' \cup R') = .940 + .605 - .590 = .955$

41. The probability of getting a 6 with one toss of a die is $\frac{1}{6}$. Therefore, in 360 tosses, we would expect to get a 6 about $\frac{1}{6}(360) = 60$ times.

45. The probability of getting 4 tails with one toss of four coins is $\frac{1}{16}$. Therefore, in 144 tosses of four coins, we would expect to get four tails about $\frac{1}{16}(144) = 9$ times.

49. The probability of rolling a sum of 2 or 12 is $\frac{1}{18}$. The probability of rolling a sum of 3 or 11 is $\frac{1}{9}$. The probability of rolling a sum of 4 or

10 is $\frac{1}{6}$. Thus, the probability of anything else happening is $1-(\frac{1}{18}+\frac{1}{9}+\frac{1}{6}) = 1-(\frac{1}{3}) = \frac{2}{3}$. So your mathematical expectation is

$$E_v = 5(\frac{1}{18}) + 2(\frac{1}{9}) + 1(\frac{1}{6}) - 1(\frac{2}{3}) = \frac{12}{18} - \frac{12}{18} = 0.$$

Yes, it is a fair game.

53. $E_v = .7(20000) - .3(10000) = \$11000$

57. The probability of getting three heads with a toss of three coins is $\frac{1}{8}$. Therefore, the odds in favor of getting three heads are 1 to 7.

61. The probability of getting a sum of 5 is $\frac{1}{9}$. Therefore, the odds in favor of getting a sum of 5 are 1 to 8.

65. If $P(E) = \frac{4}{7}$, then odds in favor are 4 to 3.

69. If odds against are 5 to 2, then odds in favor are 2 to 5. Therefore, the probability of it happening is $\frac{2}{7}$.

Problem Set 11.5

1. Let E be the event of a 5 and F the event of an odd number. Therefore, $E = \{5\}$ and $F = \{1,3,5\}$, from which we obtain $E \cap F = \{5\}$. Thus,

$$P(E|F) = \frac{P(E \cap F)}{P(F)} = \frac{\frac{1}{6}}{\frac{3}{6}} = \frac{1}{3}.$$

5. Let E be the event of a jack and F the event of a face card. Then $E \cap F = E$. Therefore,

$$P(E|F) = \frac{P(E \cap F)}{P(F)} = \frac{\frac{1}{13}}{\frac{3}{13}} = \frac{1}{3}$$

9. $$P(M|H) = \frac{P(M \cap H)}{P(H)} = \frac{.2}{.3} = \frac{2}{3}$$

$$P(H|M) = \frac{P(H \cap M)}{P(M)} = \frac{.2}{.7} = \frac{2}{7}$$

13. $$P(A|M) = \frac{P(A \cap M)}{P(M)} = \frac{\frac{5}{100}}{\frac{75}{100}} = \frac{5}{75} = \frac{1}{15}$$

$$P(F|A) = \frac{P(F \cap A)}{P(A)} = \frac{\frac{2}{100}}{\frac{7}{100}} = \frac{2}{7}$$

17. Let E = {(T,T,T), (H,T,T), (T,H,T), (T,T,H)} and F = {(H,T,T), (T,H,T), (T,T,H), (T,H,H), (H,T,H), (H,H,T)}. Therefore, $E \cap F$ = {(H,T,T), (T,H,T), (T,T,H)}.

$$P(E) = \frac{1}{2}, \quad P(F) = \frac{3}{4}, \quad P(E \cap F) = \frac{3}{8}$$

Since $P(E \cap F) = P(E) \cdot (P(F)$, the events are independent.

21. The probability of getting a double with one toss is $\frac{6}{36} = \frac{1}{6}$. Therefore, the probability of getting three successive doubles in three tosses is

$$\frac{1}{6} \cdot \frac{1}{6} \cdot \frac{1}{6} = \frac{1}{216}.$$

25. We could get a spade on the first draw and a diamond on the second draw, or vice versa. Therefore, the requested probability is

$$\frac{13}{52} \cdot \frac{13}{51} + \frac{13}{52} \cdot \frac{13}{51} = \frac{1}{4} \cdot \frac{13}{51} + \frac{1}{4} \cdot \frac{13}{51} = \frac{26}{204} = \frac{13}{102}.$$

29. $\frac{1}{52} \cdot \frac{1}{52} + \frac{1}{52} \cdot \frac{1}{52} = \frac{2}{52 \cdot 52} = \frac{1}{26 \cdot 52} = \frac{1}{1352}$

(ace of spades, king of spades, king of spades, ace of spades)

33. $\frac{5}{9} \cdot \frac{5}{9} = \frac{25}{81}$

37. $\frac{5}{13} \cdot \frac{5}{13} = \frac{25}{169}$

41. $\frac{1}{3} \cdot \frac{2}{2} + \frac{2}{3} \cdot \frac{1}{2} = \frac{1}{3} + \frac{1}{3} = \frac{2}{3}$

(red, white, white, red)

45. $\frac{5}{17} \cdot \frac{4}{16} = \frac{5}{17} \cdot \frac{1}{4} = \frac{5}{68}$

49. $\frac{3}{9} \cdot \frac{2}{8} = \frac{1}{3} \cdot \frac{1}{4} = \frac{1}{12}$

53. $\frac{1}{9} \cdot \frac{1}{9} \cdot \frac{1}{9} = \frac{1}{729}$

57. $\frac{4}{7} \cdot \frac{3}{6} \cdot \frac{2}{5} = \frac{4}{7} \cdot \frac{1}{2} \cdot \frac{2}{5} = \frac{4}{35}$

61. $\frac{4}{7} \cdot \frac{1}{3} = \frac{4}{21}$; $\quad \frac{3}{7} \cdot \frac{2}{3} = \frac{2}{7}$; $\quad \frac{3}{7} \cdot \frac{1}{3} + \frac{4}{7} \cdot \frac{2}{3} = \frac{11}{21}$

Problem Set 11.6

1. $$(x+y)^8 = x^8 + \binom{8}{1}x^7y + \binom{8}{2}x^6y^2 + \binom{8}{3}x^5y^3 + \binom{8}{4}x^4y^4 + \binom{8}{5}x^3y^5 + \binom{8}{6}x^2y^6 + \binom{8}{7}xy^7 + \binom{8}{8}y^8$$
$$= x^8+8x^7y+28x^6y^2+56x^5y^3+70x^4y^4+56x^3y^5+28x^2y^6+8xy^7+y^8$$

5. $$(a+2b)^4 = a^4 + \binom{4}{1}a^3(2b) + \binom{4}{2}a^2(2b)^2 + \binom{4}{3}a(2b)^3 + \binom{4}{4}(2b)^4$$
$$= a^4+8a^3b+24a^2b^2+32ab^3+16b^4$$

9. $$(2a-3b)^4 = (2a)^4 + \binom{4}{1}(2a)^3(-3b)^1 + \binom{4}{2}(2a)^2(-3b)^2 + \binom{4}{3}(2a)(-3b)^3 + \binom{4}{4}(-3b)^4$$
$$= 16a^4-96a^3b+216a^2b^2-216ab^3+81b^4$$

13. $(2x^2-y^2)^4 = (2x^2)^4+\binom{4}{1}(2x^2)^3(-y^2)^1+\binom{4}{2}(2x^2)^2(-y^2)^2+\binom{4}{3}(2x^2)^1(-y^2)^3+\binom{4}{4}(-y^2)^4$

$= 16x^8-32x^6y^2+24x^4y^4-8x^2y^6+y^8$

17. $(x-1)^9 = x^9+\binom{9}{1}x^8(-1)+\binom{9}{2}x^7(-1)^2+\binom{9}{3}x^6(-1)^3+\binom{9}{4}x^5(-1)^4+\binom{9}{5}x^4(-1)^5$

$+\binom{9}{6}x^3(-1)^6+\binom{9}{7}x^2(-1)^7+\binom{9}{8}x(-1)^8+\binom{9}{9}(-1)^9$

$= x^9-9x^8+36x^7-84x^6+126x^5-126x^4+84x^3-36x^2+9x-1$

21. $(a-\frac{1}{n})^6 = a^6+\binom{6}{1}a^5(-\frac{1}{n})+\binom{6}{2}a^4(-\frac{1}{n})^2+\binom{6}{3}a^3(-\frac{1}{n})^3+\binom{6}{4}a^2(-\frac{1}{n})^4$

$+\binom{6}{5}a(-\frac{1}{n})^5+\binom{6}{6}(-\frac{1}{n})^6$

$= a^6-\frac{6a^5}{n}+\frac{15a^4}{n^2}-\frac{20a^3}{n^3}+\frac{15a^2}{n^4}-\frac{6a}{n^5}+\frac{1}{n^6}$

25. $(3-\sqrt{2})^5 = (3)^5+\binom{5}{1}(3)^4(-\sqrt{2})+\binom{5}{2}(3)^3(-\sqrt{2})^2+\binom{5}{3}(3)^2(-\sqrt{2})^3$

$+\binom{5}{4}(3)(-\sqrt{2})^4+\binom{5}{5}(-\sqrt{2})^5$

$= 243-405\sqrt{2}+540-180\sqrt{2}+60-4\sqrt{2} = 843-589\sqrt{2}$

29. $(x-y)^{20} = x^{20}+\binom{20}{1}x^{19}(-y)+\binom{20}{2}x^{18}(-y)^2+\binom{20}{3}x^{17}(-y)^3+\ldots$

$= x^{20}-20x^{19}y+190x^{18}y^2-1140x^{17}y^3+\ldots$

33. $(a+\frac{1}{n})^9 = a^9+\binom{9}{1}a^8(\frac{1}{n})+\binom{9}{2}a^7(\frac{1}{n})^2+\binom{9}{3}a^6(\frac{1}{n})^3+\ldots = a^9+\frac{9a^8}{n}+\frac{36a^7}{n^2}+\frac{84a^6}{n^3}+\ldots$

37. The 4th term is

$\binom{8}{3}x^5y^3 = 56x^5y^3.$

41. The 6th term is

$\binom{7}{5}(3a)^2(b)^5 = 189a^2b^5.$

45. The 7th term is

$\binom{15}{6}(1)^9(-\frac{1}{n})^6 = \frac{5005}{n^6}.$

49. $(2-i)^6 = 2^6+\binom{6}{1}(2)^5(-i)+\binom{6}{2}(2)^4(-i)^2+\binom{6}{3}(2)^3(-i)^3$

$+\binom{6}{4}(2)^2(-i)^4+\binom{6}{5}(2)(-i)^5+\binom{6}{6}(-i)^6$

$= 64-192i-240+160i+60-12i-1$

$= -117-44i$